DIFFUSION

IN SOLIDS

SECOND EDITION

Paul Shewmon, Department of Metallurgical Engineering
Ohio State University

A Publication of The Minerals, Metals & Materials Society
420 Commonwealth Drive
Warrendale, Pennsylvania 15086
(412) 776-9024

Library of Congress Catalog Number 89-61758
ISBN Number 978-0-87339-105-4

Minerals • Metals • Materials

© 1989

PREFACE

FIRST EDITION

There are two main reasons for studying diffusion in solids. First a knowledge of diffusion is basic to an understanding of the changes that occur in solids at high temperatures. Thus it is essential that the person interested in the kinetics of precipitation, oxidation, creep, annealing, etc. be acquainted with the fundamentals of diffusion. It is often the case that a more detailed study of these processes requires the use of the more advanced concepts of diffusion in solids.

The second reason for studying diffusion is to learn more about how atoms move in solids. This is intimately connected with the study of defects in solids. Point defects are the simplest type of defect in a solid, yet their concentration and movement cannot be observed directly. Since the number of such defects and their movement can be theoretically related to the diffusion coefficient, diffusion experiments have come to be the most frequently used means of studying point defects in solids. Dislocations and grain boundaries are other types of defects, and the fact that diffusion is faster along these defects has provided a useful tool for gaining information on how atoms move in these defects.

The purpose of this book is to present a clear, concise, and relatively complete treatment of diffusion in solids. Throughout, the primary aim is to make clear the physical meaning and implications of the concepts which apply to diffusion in all crystalline solids. Where there was a choice of illustrative material, the author's background biased him toward choosing examples from metallic systems. However, those effects which are unique to nonmetallic systems are also treated in detail. Consistent with this emphasis on brevity and broad concepts, experimental results are quoted only in so far as they aid the reader in understanding the subject. Similarly, controversial issues have been avoided in favor of detailed discussion of what are felt to be the more firmly established fundamentals of the field.

3

The book has evolved from a one-semester course taught to first year graduate students and should be suitable for use at the senior or graduate student level in a course on diffusion or in a course where diffusion is discussed as a basis for studying reactions in solids. A set of problems has been included at the end of each chapter as an additional means of helping the student come to grips with the material developed in the text and apply it to some related but novel situation. Answers for many of these problems are also given.

The book should be helpful to the research worker who would like to learn more about the subject as a whole or wants a more detailed development of some topic than can be found in current papers on the subject. The chapters are largely independent of one another and the reader interested in only one topic will usually be able to fulfill his needs by reading all or parts of the chapter involved without reference to the preceding chapters.

SECOND EDITION

For the second edition the text has been largely rewritten, though the outline and purpose are the same. Important changes include:
- expansion of the treatment of diffusion in non-metals, and along surfaces
- addition of more tables of data
- more problems, along with answers for many of them.
- discussion of anomalous diffusion in various systems
- a general updating of data and discussion of the models in use.

CONTENTS

1
DIFFUSION EQUATIONS

Changes in the structure of metals and their relation to physical and mechanical properties are the primary interest of the physical metallurgist. Since most changes in structure occur by diffusion, any real understanding of phase changes, homogenization, spheroidization, etc., must be based on a knowledge of diffusion. These kinetic processes can be treated by assuming that the metal is a continuum, that is, by ignoring the atomic structure of the solid. The problem then becomes one of obtaining and solving an appropriate differential equation. In this first chapter the basic differential equations for diffusion are given, along with their solutions for the simpler boundary conditions. The diffusion coefficient is also defined, and its experimental determination is discussed.

At no point in this chapter does the atomic nature of the material enter the problem. This is not meant to detract from the importance of the study of atomic mechanisms in diffusion, since the most interesting and most active areas of study in diffusion are, and will continue to be, concerned with the information that diffusion studies can contribute to the atomic models of solids. We initially omit atomic models of diffusion to emphasize the types of problems that can be treated in this manner. In any theoretical development there are certain advantages and disadvantages to making the fewest possible assumptions. One advantage is that the results are quite generally applicable; a disadvantage is that the results are devoid of information about the atomic mechanism of the process. (Thermodynamics is an excellent example of this type of approach.) The assumptions we make in Chap. 1 were first applied to the problem in 1855 by Adolf Fick. It is indicative of the power of this approach that all the subsequent developments in the theory of solids have in no way affected the validity of the approach.

In the second and subsequent chapters we discuss the atomic processes involved in diffusion. In those chapters we present the basic differential equations for diffusion and then develop several solutions, giving examples of the application of each. The aim is to give the reader a feeling for the properties of the solutions to the diffusion equation and to acquaint him with those most frequently encountered. Thus, no attempt is made at completeness or rigor.

1.1 FLUX EQUATION

If an inhomogeneous single-phase alloy is annealed, matter will flow in a manner which will decrease the concentration gradients. If the specimen is annealed long enough, it will become homogeneous and the net flow of matter will cease. Given the problem of obtaining a flux equation for this kind of a system, it would be reasonable to take the flux across a given plane to be proportional to the concentration gradient across that plane. For example, if the x axis is taken parallel to the concentration gradient of component 1 the flux of component 1 (J_1) can be described by the equation

$$J_1 = -D_1 \left[\frac{\partial c_1}{\partial x} \right]_t \tag{1-1}$$

where D_1 is called the diffusion coefficient. This equation is called Fick's first law and fits the empirical fact that the flux goes to zero as the specimen become homogeneous, that is when the specimen reaches equilibrium. Although it need not have been the case, experiment shows that D_1, or equivalently the ratio of $-J_1$ to the concentration gradient is independent of the magnitude of the gradient. In this respect Eq. (1-1) is similar to Ohm's law, where the resistance is independent of the voltage drop, or to the basic heat-flow equations in which the conductivity is independent of the magnitude of the temperature gradient.

To emphasize the dimensions of the terms, Eq. (1-1) is written again below with the dimensions of each term given in parentheses.

$$J(\text{mass}/L^2t) = -D(L^2/t)(\partial c/\partial x)(\text{mass}/L^3)/L$$

The concentration can be given in a variety of units, but the flux must be put in consistent units. The diffusion coefficient has usually been given in units of square centimeters per second, though in the new SI units it is in square meters per second.

In a lattice with cubic symmetry, D has the same value in all directions, that is, the alloy is said to be isotropic in D. The assumption of isotropy will be made throughout the book unless a statement is made to the contrary. If there are other types of gradients in the system, other terms are added to the flux equation. These effects are interesting but complicated. They will be considered in Secs. 1-5 and 4-3.

As an example of the application of Eq. (1-1), consider the following experiment performed by Smith.[1] A hollow cylinder of iron is held in the isothermal part of a furnace. A carburizing gas is passed though the inside of the cylinder, and a decarburizing gas over the outside. When the carbon concentration at each point in the cylinder no longer changes with time, that is $(\partial c/\partial t) = 0$ throughout, the quantity of carbon passing though the cylinder per unit time (q/t) is a constant. However, since J is the flow per unit area, it is a function of the radius r and is given by the equation

$$J = q/At = q/2\pi rlt \qquad (1\text{-}2)$$

where l is the length of the cylinder through which carbon diffusion occurs. Combining Eqs. (1-1) and (1-2) gives an equation for q, the total amount of carbon which passed through the cylinder during the time t:

$$q = -D(2\pi lt)\, dc/d(lnr) \qquad (1\text{-}3)$$

For a given run, q, l, and t can be measured. If the carbon concentration through the cylinder wall is determined by chemical analysis, D can be determined from a plot of c versus lnr. Such a plot will be a straight line if the diffusion coefficient does not vary with composition. However, for carbon in γ-iron Smith found that the slope of this plot $(dc/dlnr)$ became smaller in passing from the low-carbon side of the tube to the high-carbon side. An example of his results for 1000° C is shown in Fig. 1-1. At this temperature the diffusion coefficient varies from 2.5×10^{-7} cm^2/sec at 0.15 weight per cent carbon to 7.7×10^{-7} cm^2/sec at 1.4 weight per cent carbon.

Similar experiments have frequently been performed by passing a gas though a membrane. Often the membranes are so thin it is impossible to determine the concentration as a function of distance in the membrane by means of chemical analysis. The experimental results therefore consist of a measured steady-state flux, the pressure drop across the membrane, and the thickness of the membrane (Δy). This flux, for a given pressure drop, is called the permeability. To obtain a value of D from these data, the value of $\partial c/\partial y$ inside the membrane must be determined. One way to do this is to assume that the value of c in the metal at each gas-metal interface is the value that would exist in equilibrium with the gas if there were no net flux. This would

---

[1]R. P. Smith, *Acta Met.*, *1* (1953) 578.

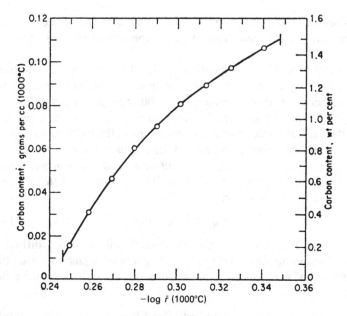

Fig. 1-1—c vs. log r for a hollow cylinder of iron after attaining a steady state with a carburizing gas passing through the inside and a decarburizing gas passing over the outside (1000° C). [R. P. Smith, *Acta Met., 1* (1953) 578.]

be true if the solution of gas into the surface of the metal occurred much more rapidly than the diffusion out of the surface region into the rest of the metal. Experimentally, this assumption is checked by determining the fluxes for two thicknesses of membrane under the same pressure differential and at the same temperature. If equilibrium does exist at the gas-metal interface, then Δc is the same for both cases, and from Eq. (1-1)

$$J = -D \, \Delta c / \Delta y$$

that is, J will be inversely proportional to Δy and both thicknesses will give the same value of D. At the other extreme, if the rate of solution of gas at the interface determines the flux, the flux will be the same for both membrane thicknesses, and the value of D in the membrane cannot be obtained from the flux.

1.2   DIFFUSION EQUATION

If a steady state does not exist, that is, if the concentration at some point is changing with time, Eq. (1-1) is still valid, but it is not a convenient form to use. To obtain more useful equations, it is necessary to start with a second differential equation. It is obtained by using Eq. (1-1) and a material balance. Consider a bar of unit cross-sectional area with the x axis along its center. An element Δx thick along the x axis has flux J_1 in one side and J_2 out the other (see Fig. 1-2). If Δx is very small J_1 can be accurately related to J_2 by the expression

$$J_1 = J_2 - \Delta x(\partial J/\partial x) \qquad (1-4)$$

Since the amount of material that came into the element in unit time (J_1) is different from that which left (J_2), the concentration in the ele-

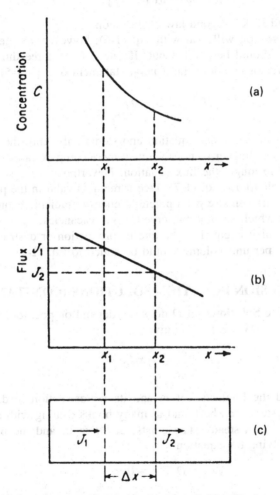

Fig. 1-2—(a) shows an assumed $c(x)$ plot, (b) shows $J(x)$ for this plot, and (c) shows the element of volume with the flux J_1 entering and J_2 leaving.

ment has changed. The volume of the element is $1 \cdot \Delta x$ (unit area times the thickness), so the net increase in matter in the element can be expressed by any part of the equation

$$J_1 - J_2 = \Delta x(\partial c/\partial t) = -\Delta x(\partial J/\partial x) \tag{1-5}$$

Now Eq. (1-1) is valid at any instant even if the concentration and concentration gradient at that point are changing with time. Therefore we can substitute it into Eq. (1-5).

$$\frac{\partial c}{\partial t} = \frac{\partial}{\partial x}\left[D\,\frac{\partial c}{\partial x}\right] \tag{1-6}$$

This is called Fick's second law of diffusion.

The next section will deal with Eq. (1-6); however, the generality of Eq. (1-6) should be pointed out. If one goes to three dimensions and uses a vector notation, the general statement of Eq. (1-5) is

$$\frac{\partial c}{\partial t} = -\nabla J \tag{1-7}$$

This is called a continuity equation and stems only from the conservation of matter. In later sections we treat more complex cases in which Eq. (1-1) is no longer the flux equation. Nevertheless, we shall continue to be able to use Eq. (1-7) since it remains valid in the presence of additional gradients, e.g., a potential-energy gradient. If one deals with entities which are not conserved, e.g., vacancies, then an additional term which is equal to the rate of production or destruction of these entities per unit volume should be added to Eq. (1-7).

1.3 DIFFUSION EQUATION SOLUTIONS (CONSTANT D)

Steady-state Solutions. If D does not depend on position, and we can take $\mathbf{J} = -D\nabla c$, Eq. (1-7) gives

$$\frac{\partial c}{\partial t} = D\,\nabla^2 c \tag{1-8}$$

$\nabla^2 c$ is called the Laplacian of c, and its representation in different coordinate systems can be found in many books dealing with applied mathematics.[2] If a steady state exists, $\partial c/\partial t = 0$, and the problem reduces to solving the equation

$$D\,\nabla^2 c = 0 \tag{1-9}$$

---

[2] J. Crank, *Mathematics of Diffusion*, 2nd ed., Oxford Univ. Press, 1975.

The simplest cases are, for cartesian coordinates in one dimension,

$$D \frac{\partial^2 c}{\partial x^2} = 0 \qquad (1\text{-}10)$$

for cylindrical coordinates where $c = f(r)$,

$$\frac{D}{r} \frac{\partial}{\partial r} \left[r \frac{\partial c}{\partial r} \right] = D \left[\frac{\partial^2 c}{\partial r^2} + \frac{1}{r} \frac{\partial c}{\partial r} \right] = 0 \qquad (1\text{-}11)$$

and, for spherical coordinates where $c = f(r)$,

$$\frac{D}{r^2} \frac{\partial}{\partial r} \left[r^2 \frac{\partial c}{\partial r} \right] = D \left[\frac{\partial^2 c}{\partial r^2} + \frac{2}{r} \frac{\partial c}{\partial r} \right] = 0 \qquad (1\text{-}12)$$

We shall not deal with the solutions to these equations except as they arise in a few examples below.

The differential equation represented by Eq. (1-9) arises in many branches of physics and engineering. In two or more dimensions the solutions can be quite complicated. The interested reader should consult books on heat flow,[3] or potential theory.[4]

Non-steady-state Solutions. If D is not a function of position, i.e., composition, Eq. (1-6) becomes

$$\frac{\partial c}{\partial t} = D \frac{\partial^2 c}{\partial x^2} \qquad (1\text{-}13)$$

We wish to determine the concentration as a function of position and time, that is, $c(x,t)$, for a few simple initial and boundary conditions. In general, the solutions of Eq. (1-13) for constant D fall into two forms. When the diffusion distance is short relative to the dimensions of the initial inhomogeneity, $c(x,t)$ can be most simply expressed in terms of error functions. When complete homogenization is approached, $c(x,t)$ can be represented by the first few terms of an infinite trigonometric series. (In the case of a cylinder, the trigonometric series is replaced by a series of Bessel functions.) The reader interested in a comprehensive listing of solutions should consult Crank or Carslaw and Jaeger.[3]

Thin-film Solution. Imagine that a film only b in thickness with solute concentration c_o is plated on one end of a long rod of solute-free material. If a similar solute-free rod is welded to the plated end of this rod (without any diffusion occurring) and the rod is then an-

[3]H. S. Carslaw, J. C. Jaeger, *Conduction of Heat in Solids*, Oxford Univ. Press, 1959.
[4]O. D. Kellog, *Foundations of Potential Theory*, Springer-Verlag, New York, 1967.

nealed for a time t so that diffusion can occur, the concentration of solute along the bar will be given by the equation

$$c(x,t) = \frac{bc_o}{2\sqrt{\pi Dt}} \exp\left[\frac{-x^2}{4Dt}\right] \qquad (1\text{-}14)$$

in those regions where $\sqrt{Dt} > b$.[5] Here x is the distance in either direction normal to the initial solute film. To show that Eq. (1-14) is the correct solution, two steps are necessary. First, differentiation shows that it is indeed a solution to Eq. (1-13). Second, the equation satisfies the boundary conditions of the problem since

$$\text{for} \quad |x| > 0, \quad c \to 0 \quad \text{as} \quad t \to \infty$$
$$\text{for} \quad x = 0, \quad c \to \infty \quad \text{as} \quad t \to 0$$

yet the total quantity of solute is fixed since

$$\int_{-\infty}^{\infty} c(x,t)\, dx = bc_o$$

The characteristics of this solution can best be seen with the help of Fig. 1-3. Here the concentration is plotted against distance after some diffusion has occurred. As more diffusion occurs, the $c(x)$ curve will spread out along the x axis. However, since the amount of solute is fixed, the area under the curve remains fixed. To understand how this occurs, observe that $c(x = 0)$ decreases as $1/\sqrt{t}$ while the distance between the plane $x = 0$ and the plane at which c is $1/e$ times $c(x = 0)$ increases as $\sqrt{t}$. This distance is given by the equation $x = 2\sqrt{Dt}$.

In Fig. 1-3b is plotted dc/dx versus x. This is proportional to the flux across any plane of constant x. It will be seen that it goes to zero at $x = 0$ and at large positive or negative values of x.

In Fig. 1-3c is plotted d^2c/dx^2 versus x. This quantity is proportional to the rate of accumulation of solute in the region of any plane of constant x. It is also proportional to the curvature of the $c(x)$ plot. Thus it is seen that in the region around $x = 0$, $c(x)$ is concave downward and the region is losing solute. The concave upward regions on the outer portions of the $c(x)$ curve are gaining solute; regions at large values of x are undergoing no change in solute content. To develop a feeling for these curves, the student is urged to derive the latter two for himself by plotting the slope of the curve above it versus x.

Equation (1-14) is often referred to as the solution for a thin film in

---

[5]For thick films, the right side of Eq. (1-14) must be multiplied by the term $(1 - b^2/(12Dt))$ [W. A. Johnson, *Trans. AIME*, *147* (1942) 331.] With the carrier free radioactive tracers now available, this correction is almost never necessary.

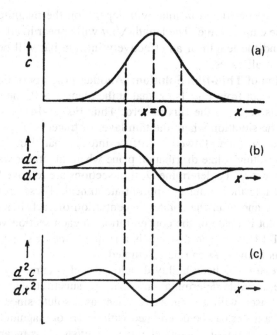

Fig. 1-3 — (a), (b), and (c) show $c(x)$, dc/dx, and d^2c/dx^2 versus x, for Eq. (1-14), in arbitrary units.

the middle of an "infinite bar." Since no bar is truly infinite, it is of value to consider just how long a bar must be for this equation to apply. If the thin film is placed in the middle of a short bar and none of the solute is lost when it reaches an end of the bar, that solute which would normally have diffused past the end will be reflected back into the specimen, and c in that region will be higher than given by Eq. (1-14). Thus a short bar can be considered infinite if the quantity of solute which would lie outside its length in a truly infinite bar is an insignificant portion of the total solute present. Arbitrarily taking 0.1% as a sufficiently insignificant portion, we need to solve for x' in the equation

$$0.001 = (2/bc_o) \int_{x'}^{\infty} c(x,t)\, dx$$

where $(bc_o/2)$ is the total quantity of solute in half of the bar and the integral is the quantity beyond x'.

The solution to this equation is $x' = 4.6\sqrt{Dt}$. As might have been expected, the answer is stated in terms of the quantity $\sqrt{Dt}$. For sufficiently short times any bar is "infinite," and the time during which the bar can be considered infinite will depend on the magnitude of D as well as the elapsed time. The length $\sqrt{Dt}$ will appear in all diffusion problems, and the length of an effectively infinite bar will be several times $\sqrt{Dt}$ in all cases.

Application of Thin-film Solution. Another property of Eq. (1-14) which is apparent from Fig. 1-3 is that at the plane $x = 0$, the gradient, dc/dx, equals zero, so the flux is zero. Thus Eq. (1-14) can be used to describe the situation where the thin layer of tracer is placed on one end or a bar and then allowed to diffuse into the bar. Eq. (1-14) describes the resulting solute distribution plane where the tracer was placed is defined as $x = 0$. To determine D, thin sections are removed parallel to the initial interface, after an appropriate anneal. These are sections of constant x, and after the solute concentration of each is measured, a semi-log plot is made of the concentration in each section versus x^2. From Eq. (1-14) it is seen that this is a straight line of slope $-1/4Dt$ so that if t is known, D can be calculated.

This procedure has been highly developed and is currently used for all the more accurate determinations of D for substitutional atoms. It is invariably used with a radioactive tracer as a solute since the concentration of a tracer can be determined with orders of magnitude greater sensitivity than is possible using chemical analysis. This means that D can be measured with extremely small concentration changes. One of the other advantages of using radioactive tracers is that it is just as easy to study the diffusion of a silver tracer in silver as it is to study the diffusion of cadmium tracer in silver. In both cases there is concentration gradient for the tracer atom involved, and there will therefore be a spreading out of the tracer with time. The fact that the tracer is chemically very similar to the solvent in one case makes no difference in the application of the diffusion equations.

To help the reader develop an understanding of the magnitude of the values of D in metals and the procedures involved in their measurement, let us go through a rough calculation of the values of D which can be determined with this type of experiment. For the case of substitutional atoms in metals the value of D at the melting point is usually about 10^{-8} cm^2/s, so that this sets a rough upper limit on the values of D to be measured with this technique. If an accurate value of the slope of ln c versus x^2 is to be obtained, it is necessary to have several sections, say ten. If the ratio of maximum to minimum concentration is 1000 this corresponds to a value of $x^2/4Dt$ of about 7. Setting an upper limit on the diffusion time of 10^6 s (12 days), the minimum value of D that can be measured depends primarily on the thickness of the sections that can be removed and collected. If this is done with a lathe, the minimum section is about 0.001 cm and the corresponding minimum D is about 10^{-13} cm^2/s. Thus D can be easily measured over five orders of magnitude. This may at first seem like an appreciable range, but because D varies rapidly with temperature, it allows the determination of D from the melting point down to only about 0.7 of the absolute melting temperature; most diffusion controlled reactions of interest in solids occur at temperatures lower than this.

Micro-sectioning techniques have been developed that allow sections of 3 nm to be taken (by plasma sputtering or oxide stripping techniques). This reduction in section size means that values of D down to 10^{-20} or 10^{-21} cm^2/s can be measured. The reader particularly interested in the experimental determination of D can consult the review article by Rothman.[6]

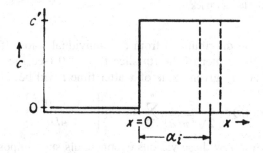

Fig. 1-4

Solutions for a Pair of Semi-infinite Solids. Consider the initial distribution which results if a piece of pure A is joined to pure B without interdiffusion. This distribution is shown graphically in Fig. 1-4. The boundary conditions are given by

$$c = 0 \quad \text{for} \quad x < 0, \quad \text{at} \quad t = 0$$
$$c = c' \quad \text{for} \quad x > 0, \quad \text{at} \quad t = 0$$

A solution to the diffusion equation for this case can be obtained in the following manner: Imagine that the region of $x > 0$ consists of n slices, each of thickness $\Delta\alpha$ and unit cross-sectional area. Consider one particular slice. It initially contains $c'\Delta\alpha$ of solute, and if the surrounding regions were initially solute free, the distribution after some diffusion would be that given by the thin-film solution, i.e., Eq. (1-14). The fact that there is solute in the adjacent slices does not in any way affect this result, and the actual solution is thus given by a su-

---

[6]S. J. Rothman, "Measurement of Tracer Diffusion Coefficients in Solids," *Diffusion in Crystalline Solids*, ed. G. E. Murch, A. S. Nowick, Academic Press, 1984, p. 1–61.

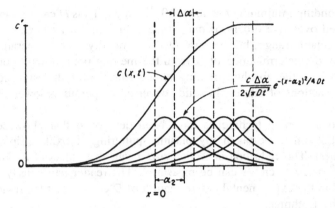

Fig. 1-5—$c(x,t)$ is the sum of the exponential curves which represent the solute diffusing out of each slab $\Delta\alpha$ thick.

perposition of the distributions from the individual slabs. If α_i is the distance from the center of the ith slice to $x = 0$ (see Fig. 1-5), the concentration at any given value of x after time t will be

$$c(x,t) = \frac{c'}{2\sqrt{(\pi Dt)}} \sum_{i=1}^{n} \Delta\alpha \exp\left[-\frac{(x - \alpha_i)^2}{4Dt} \right] \qquad (1\text{-}15)$$

Figure 1-5 shows how these various exponentials superimpose to give the actual distribution for the case of rather thick slices. In the limit of n going to infinity, $\Delta\alpha_i$ goes to zero, and from the definition of an integral

$$c(x,t) = \frac{c'}{2\sqrt{(\pi Dt)}} \int_{o}^{\infty} \exp\left[-\frac{(x - \alpha)^2}{4Dt} \right] d\alpha \qquad (1\text{-}16)$$

Substituting $(x - \alpha)/2\sqrt{Dt} = u$ we can rewrite the solution[7]

$$c(x,t) = \frac{c}{\sqrt{\pi}} \int_{-\infty}^{x/2\sqrt{Dt}} \exp(-u^2)\, du \qquad (1\text{-}17)$$

This type of integral appears quite generally in the solutions of problems where the initial source of solute is an extended one and the diffusion distance $2\sqrt{Dt}$ is small relative to the length of the system. The integral cannot be evaluated in any simple manner, but because of its frequent appearance in diffusion and heat-flow problems, its values are

---

[7]That this is a solution can be shown by differentiating Eq. (1-17) and substituting in Eq. (1-13). Differentiating an argument of an integral is discussed in most books on advanced calculus or differential equations.

Table 1-1.   The Error Function

z	erf(z)	z	erf(z)
0	0	0.85	0.7707
0.025	0.0282	0.90	0.7969
0.05	0.0564	0.95	0.8209
0.10	0.1125	1.0	0.8427
0.15	0.1680	1.1	0.8802
0.20	0.2227	1.2	0.9103
0.25	0.2763	1.3	0.9340
0.30	0.3286	1.4	0.9523
0.35	0.3794	1.5	0.9661
0.40	0.4284	1.6	0.9763
0.45	0.4755	1.7	0.9838
0.50	0.5205	1.8	0.9891
0.55	0.5633	1.9	0.9928
0.60	0.6039	2.0	0.9953
0.65	0.6420	2.2	0.9981
0.70	0.6778	2.4	0.9993
0.75	0.7112	2.6	0.9998
0.80	0.7421	2.8	0.9999

available in tabular form. The function given in Table 1-1 is called an error function and is defined by the equation[8]

$$erf(z) = \frac{2}{\sqrt{\pi}} \int_{o}^{z} \exp(-u^2)\, du \qquad (1\text{-}18)$$

It can be shown that erf(∞) = 1, and it is evident that

$$erf(-z) = -erf(z)$$

Equation (1-17) can thus be rewritten

$$c(x,t) = \frac{c'}{2}\left[1 + erf\left(\frac{x}{2\sqrt{Dt}}\right)\right] \qquad (1\text{-}19)$$

This has already been plotted in Fig. 1-5.

It should be noted that each value of the ratio c/c' is associated with a particular value of $z = x/2\sqrt{Dt}$. Thus $z = 1$ is always associated with $c/c' = 0.92$; the position of the plane whose composition is 0.92 c' is given by the equation $x = 2\sqrt{Dt}$. Further inspection shows that each composition moves away from the plane of $x = 0$ at a rate proportional to $\sqrt{Dt}$, with the exception of $c = c'/2$, which corresponds to $z = 0$ and thus remains at $x = 0$.

[8]The function erf(z) can be evaluated by numerical integration, or with the infinite series erf(z) = $(2/\sqrt{\pi})(z - z^3/3*1! + z^5/5*2! - z^7/7*3! \ldots .)$.

"Infinite" System—Surface Composition Constant. Eq. (1-19) gives a constant composition at the plane $x = 0$ independent of time. Thus Eq. (1-19) can be used in the region $x > 0$ to describe the case in which an initially homogeneous alloy of solute c' is held in an atmosphere which reduces the surface concentration to $c'/2$ and keeps it there for all $t > 0$. The boundary conditions are

$$c = c'/2 \quad \text{for} \quad x = 0, \quad \text{at} \quad t > 0$$
$$c = c' \quad \text{for} \quad x > 0, \quad \text{at} \quad t = 0$$

and the solution is still Eq. (1-19). If the surface concentration is held at $c = 0$ instead of $c'/2$ for all $t > 0$, the solution becomes

$$c(x,t) = c' \, \text{erf}\left[\frac{x}{2\sqrt{Dt}}\right] \qquad (1\text{-}20)$$

If the surface concentration of an initially solute-free specimen is maintained at some composition c'' for all $t > 0$, solute is added to the specimen, and the solution is equivalent to Eq. (1-19) in the region $x < 0$. Since $\text{erf}(-z) = -\text{erf}(z)$, the solution for this case (in the region $x > 0$) is

$$c(x,t) = c''\left[1 - \text{erf}\left(\frac{x}{2\sqrt{Dt}}\right)\right] \qquad (1\text{-}21)$$

Inspection of this equation shows that it fits the situation, since for $x = 0$, $c = c''$ and at $x \gg 2\sqrt{Dt}$, c is about equal to zero.

It should be pointed out that in any of the solutions given by Eqs. (1-14) and (1-19) to (1-21), the zero of concentration can be shifted to fit the case where the initial concentration is not zero, but some other constant value, say c_o. As an example, if the boundary conditions are

$$c = c_o \quad \text{for} \quad x > 0, \quad \text{at} \quad t = 0$$
$$c = c' \quad \text{for} \quad x < 0, \quad \text{at} \quad t = 0$$

the solution in Eq. (1-19) is changed to

$$c(x,t) - c_o = \frac{c' - c_o}{2}\left[1 - \text{erf}\left(\frac{x}{2\sqrt{Dt}}\right)\right] \qquad (1\text{-}22)$$

The assumption of a constant D independent of position in the couple places a severe limitation on the use of this type of solution in making accurate determinations of D. If D is to be measured with a tracer in a chemically homogeneous alloy, it is usually easier to use the thin-film solution discussed earlier. On the other hand, if D is to be determined in a couple which has a range of chemical compositions in it, D will usually vary with the position, i.e. composition, and the Matano-Boltzmann solution will be required (see Sec. 1-6 of this chapter).

The most frequent use of the error-function solutions arises when it is desired to estimate the amount of diffusion that will occur in a system where D is known to vary across the diffusion zone. A complete solution of the problem with a variable D is quite time-consuming, and essentially the same answer can be obtained by using an average D. This problem is found in the carburizing or decarburizing of steel, for which Eqs. (1-20) and (1-21) will often give adequate answers. Another case, in which these same equations would not give as accurate an answer, is in the dezincing of a Cu-Zn alloy. Here, as the zinc is removed, the sample shrinks, thus moving the original interface relative to the interior of the sample. No shrinkage was allowed for in the derivations of Eqs. (1-19) to (1-21); this further detracts from the accuracy of the answers obtained with these equations.

Finite Systems—Complete Homogenization. The above solutions have dealt with infinite systems. We now consider solutions for "small" systems, that is, for those which approach complete homogenization. It is first assumed that there exist solutions which are the product of a function only of time $T(t)$ and a function of distance $X(x)$. That is, we assume that

$$c(x,t) = X(x)T(t) \tag{1-23}$$

It may be noted that the solutions discussed up to this point are excluded from this family since they are of the form $c(x,t) = f(x/\sqrt{t})$.

If we differentiate Eq. (1-23) in the prescribed manner and substitute in Fick's second law, the result is

$$X \frac{dT}{dt} = DT \frac{d^2X}{dx^2}$$

or

$$\frac{1}{DT} \frac{dT}{dt} = \frac{1}{X} \frac{d^2X}{dx^2} \tag{1-24}$$

The equation now contains only total differentials. The left side is a function only of time, and the right a function only of distance. But since x and t can be varied independently, Eq. (1-24) can be satisfied only if both sides of the equation are equal to a constant. This constant will be designated as $-\lambda^2$, where λ is a real number. The differential equation in time then is

$$\frac{1}{T}\frac{dT}{dt} = -\lambda^2 D$$

which integrates to

$$T = T_o \exp(-\lambda^2 Dt) \tag{1-25}$$

where T_o is a constant. The reason for requiring that the quantity $-\lambda^2$ have only negative values stems from our desire to deal only with systems in which any inhomogeneities disappear as time passes, i.e., that T approaches zero as t increases.

The equation in x is

$$\frac{d^2X}{dx^2} + \lambda^2 X = 0$$

Since λ^2 is always positive, the solution to this equation is of the form

$$X(x) = (A' \sin \lambda x + B' \cos \lambda x) \tag{1-26}$$

where A' and B' are constants.

Combining the solutions for T and X gives

$$c(x,t) = (A \sin \lambda x + B \cos \lambda x) \exp(-\lambda^2 Dt)$$

But if this solution holds for any real value of λ, then a sum of solutions with different values of λ is also a solution. Thus in its most general form the product solution will be an infinite series of the form

$$c(x,t) = A_o + \sum_{n=1}^{\infty} (A_n \sin \lambda_n x + B_n \cos \lambda_n x) \exp(-\lambda_n^2 Dt) \tag{1-27}$$

where A_o is the average concentration after homogenization has occurred.

Diffusion out of a Slab. As an example of the use of Eq. (1-27), consider the loss of material out both sides of a slab of thickness h. The boundary conditions to be assumed are

$$c = c_o \quad \text{for} \quad 0 < x < h, \quad \text{at} \quad t = 0$$
$$c = 0 \quad \text{for} \quad x = h \quad \text{and} \quad x = 0, \quad \text{at} \quad t > 0$$

Ultimately the concentration in the slab goes to zero so $A_o = 0$. By setting all B_n equal to zero, c will be zero at $x = 0$ for all times. To make $c = 0$ at $x = h$, the argument of $\sin \lambda_n x$ must equal zero for $x = h$. This is done by letting $\lambda_n = n\pi/h$, where n is any positive integer. If we substitute $B_n = 0$ and $\lambda_n = n\pi/h$ into Eq. (1-27), the first boundary condition requires that

$$c_o = \sum_{n=1}^{\infty} A_n \sin(xn\pi/h) \tag{1-28}$$

To determine the A_n which will satisfy this equation, multiply both sides of this equation by $\sin(xp\pi/h)$, and integrate x over the range $0 < x < h$. This gives the equation

$$\int_o^h c_o \sin(xp\pi/h)\, dx = \sum_{n=1}^{\infty} A_n \int_o^h \sin(xp\pi/h) \sin(xn\pi/h)\, dx$$

Each of the infinity of integrals on the right equals zero, except the one in which $n = p$. This integral is equal to $h/2$. The values of A_n which will satisfy Eq. (1-28) are thus given by the equation

$$A_n = (2/h) \int_o^h c_o \sin(nx\pi/h)\, dx \tag{1-29}$$

Integration of this equation shows that $A_n = 0$ for all even values of n and $A_n = 4c_o/n\pi$ for odd values of n. Changing the summation index so that only odd values of n are summed gives

$$A_n = A_j = 4c_o/(2j + 1)\pi \quad j = 0, 1, 2 \ldots \tag{1-30}$$

The solution is thus

$$c(x,t) = \frac{4c_o}{\pi} \sum_{j=0}^{\infty} \frac{1}{(2j + 1)} \sin \frac{(2j + 1)\pi x}{h}$$
$$\cdot \exp\left(-\left[\frac{(2j + 1)\pi}{h}\right]^2 Dt\right) \tag{1-31}$$

A moment's study of this equation shows that each successive term is smaller than the preceding one. Also, the percentage decrease between terms increases exponentially with time. Thus after a short time has elapsed, the infinite series can be satisfactorily represented by only a few terms, and for all time beyond some period t', $c(x,t)$ is given by a single sine wave. To determine the error involved in using just the first term to represent $c(x,t)$ after time t', it is easiest to consider the ratio of the maximum values of the first and second terms. This ratio R is given by the equation

$$R = 3 \exp(8\pi^2 Dt'/h^2)$$

For $h = 4.75\sqrt{Dt}$, R is about 100, so that for $h^2 < 22\, Dt$ (or $t > 0.044h^2/D$) the error in using the first term to represent $c(x,t)$ is less than 1%.

This solution could be applied to the decarburization of a thin sheet of steel, and it is worthwhile to compare the use of this series solution with the error-function solution of Eq. (1- 20). For short times the sheet thickness can be considered infinite. The carbon distribution below each surface will then be given by the error-function solution as well as by this series solution. To evaluate $c(x,t)$ in this case using Eq. (1-31) would require the evaluation of many terms, and it is easier to look up the error function in a table. This is true until $h \simeq 3.2\sqrt{Dt}$, at which time $R \simeq 20$, and the error in using the error function is about 2% at the plane $x = h/2$. For times greater than this, the first term in Eq. (1-31) becomes a better approximation and would be used.

One of the most frequent metallurgical applications of this type of solution appears in the degassing of metals. Here it is often difficult to determine the concentration at various depths, and what is experimentally determined is the quantity of gas which has been given off or the quantity remaining in the metal. For this purpose the average concentration $\bar{c}$ is needed. This is obtained by integrating Eq. (1-31):

$$\bar{c}(t) = \frac{1}{h} \int_o^h c(x,t) \, dx$$

$$= \frac{8c_o}{\pi^2} \sum_{j=0}^{\infty} \frac{1}{(2j + 1)^2} \exp\left(-\left[\frac{(2j + 1)\pi}{h}\right]^2 Dt\right) \qquad (1\text{-}32)$$

The ratio of the first and second terms in this series is three times as large as in the case of Eq. (1-31), and for $\bar{c} < 0.8 \, c_o$ the first term is an excellent approximation to the solution. The solution for $\bar{c}/c_o < 0.8$ can be written

$$\bar{c}/c_o = 8/\pi^2 \exp(-t/\tau) \qquad (1\text{-}33)$$

where $\tau = h^2/\pi^2 D$ is called the relaxation time. Equation (1-31) is a type that is met frequently in systems that are relaxing to an equilibrium state. The quantity τ is a measure of how fast the system relaxes; when $t = \tau$, the concentration has relaxed to roughly two-thirds of its initial value. Large values of τ thus characterize slow processes.

Equations similar to (1-32) and (1-33) and derived for the degassing of cylinders have been used in the accurate measurement of D for hydrogen in nickel.[9] The form of the equations is identical, but the equation for τ varies somewhat. For a long cylinder of diameter d the relaxation time is $\tau = d^2/4.8^2 D$, while for a sphere of diameter d the equation is $\tau = d^2/4\pi^2 D$. Noting that the maximum dimension of the cylinder and sphere, d, is comparable to the thickness of the plate, h, it is seen that as the surface to volume ratio of the solid increases, the relaxation time gets shorter, for a given value of D.

---

[9]M. Hill, E. Johnson, Acta Met., 3 (1955) 566.

1.4 KINETICS OF PRECIPITATION

We consider next the solution of a much more complicated problem, namely, the kinetics of the removal of solute from a supersaturated matrix by the growth of randomly distributed precipitate particles. The average composition of the solute in solution, $\bar{c}(t)$, can be easily and continuously measured by several techniques. Thus the problem is to determine the relationship between the time variation of $\bar{c}(t)$ and the diffusion coefficient, particle shape, average interparticle distance, or other parameters that may be determined from independent observations or may need to be determined.

The problem is very complicated, but the complication comes primarily from the number of particles rather than the complexity of the diffusion around each particle. Thus the first task is to make a series of simplifying assumptions which make the problem tractable without making it so idealized that it bears no relation to the experimental system. A detailed mathematical analysis of the problem and the errors resulting from the different approximations has been given by Ham.[10]

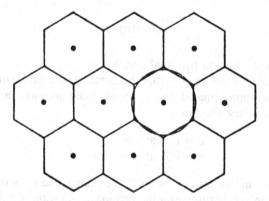

Fig. 1-6—Close-packed plane out of close-packed lattice of ppt. particles, and traces of planes midway between particles. Circle is trace of equivalent sphere used in analysis.

Simplifications. Since the precipitate particles are randomly distributed it is not unreasonable to approximate their distribution by that of a face-centered cubic (close-packed) space lattice. If planes are passed midway between all nearest-neighbor particles, the planes enclose each particle in a separate cell. [See Fig. 1-6 for a cut along the (111) plane.] Each of these planes is a plane of mirror symmetry. There is no net flux across a symmetry plane, since in the absence of a source or sink in the symmetry plane, the only way it can be a symmetry plane is

[10]F. S. Ham, *J. Phys. Chem. Solids*, 6 (1958) 335–51.

for ∇c to be zero across it. Thus there is no net flux into or out of each cell. Each cell can thus be treated as if its walls were impermeable to solute. Determining $\bar{c}(t)$ for the aggregate is thus reduced to determining $\bar{c}(t)$ for only one cell. The mathematical description of the solution can be still further simplified without significant loss of accuracy if the cell bounded by segments of planes is replaced by a sphere of equivalent volume. The radius of the equivalent sphere is defined as r_e. We are now ready to proceed to the solution of the problem.

Solution for Short Time. Consider first the initial period of precipitation when the solute-drained region is small relative to the size of the equivalent sphere, that is, $r_e \gg \sqrt{Dt}$. The first equation is obtained from the fact that $\bar{c}$ decreases as solute atoms leave solution and precipitate on the precipitate surface. The amount of material leaving solution per unit time can be expressed both as the flux into the precipitate times the surface area of the precipitate and as the volume of the cell times the rate of change of $\bar{c}$ in the cell. If the precipitate is taken to be a sphere of radius α, the equality of these terms gives the equation

$$\frac{4\pi r_e^3}{3} \frac{d\bar{c}}{dt} = J(\alpha)4\pi\alpha^2$$

where $J(\alpha)$ is the flux in the solid solution at $r = \alpha$.

To evaluate $J(\alpha)$ we assume that the actual solute distribution near $r = \alpha$ can be approximated by the steady-state solution that should satisfy the boundary conditions

$$c = c_o \quad \text{at} \quad r = r_e$$
$$c = c' \quad \text{at} \quad r = \alpha(t)$$

where c_o is the initial concentration in the matrix and c' is the matrix concentration in equilibrium with the precipitate. Obviously this steady-state solution will not be obeyed throughout the equivalent sphere, but it is to be used only at $r = \alpha$ and then only for $\partial c/\partial r$ at $r = \alpha$. Letting α vary with time, the value of $\partial c/\partial r$ determined from this solution has the correct time dependence, and indeed for very dilute solutions the error involved in its use is insignificant.

In spherical coordinates at the steady-state, Fick's second law gives

$$\frac{\partial c^2}{\partial r^2} + \frac{2}{r} \frac{\partial c}{\partial r} = 0 \qquad (1\text{-}12)$$

The required solution is

$$c = b/r + d$$

Using the given boundary conditions, and assuming $r_e \gg \alpha(t)$ we have

$$c(r) = -(c_o - c')\alpha/r + c_o$$

Thus

$$J(\alpha) = -D\left(\frac{\partial c}{\partial r}\right)_{r=\alpha(t)} = -D\frac{(c_o - c')}{\alpha(t)}$$

or

$$\frac{d\bar{c}}{dt} = -\frac{3D}{r_e^3}(c_o - c')\alpha(t) \tag{1-34}$$

A second equation can be obtained from the conservation of solute. Defining c'' as the concentration per unit volume of solute in the precipitate and assuming $\alpha = 0$ at $t = 0$, we have

$$(4/3)\pi c''\alpha^3(t) = (4/3)\pi r^3[c_o - \bar{c}(t)]$$

Solving this equation for $\alpha(t)$ gives

$$\alpha(t) = r_e[(c_o - \bar{c}(t))/c'']^{1/3}$$

Substituting this in Eq. (1-34) gives the desired differential equation

$$\frac{d\bar{c}}{dt} = \frac{-3D(c_o - c')}{c''^{1/3}r_e^2}[c_o - \bar{c}]^{1/3} \tag{1-35}$$

Defining $b = (3D/r_e^2)(c_o - c')/c''^{1/3}$, Eq. (1-35) becomes

$$d\bar{c}/dt = -b(c_o - \bar{c})^{1/3}$$

This integrates to

$$-(3/2)(c_o - \bar{c})^{2/3} = -bt + \beta$$

but at $t = 0$, c_o must equal $\bar{c}$, so $\beta = 0$. Thus

$$\bar{c} = c_o - (2bt/3)^{3/2} \tag{1-36}$$

This equation then gives the solution for short times, that is $\sqrt{Dt} \ll r_e$. Ham shows that if $\alpha(0) = 0$, Eq. (1-36) is a good approximation for $\bar{c}/c_o > 2/3$. He also gives other solutions which are valid for more complete precipitation, that is, $\bar{c}/c_o < 2/3$, as well as a discussion of the case in which $\alpha \neq 0$.

Let us now examine the properties of this solution. First, it is seen that for a spherical precipitate $(c_o - \bar{c})$ is proportional to $t^{3/2}$. This stems from the fact that the radius of the region drained of solute by the precipitate particle is proportional to $t^{1/3}$. Equation (1-36) simply states that $\bar{c}$ equals c_o minus a term proportional to the quantity of solute precipitated. For spherical particles the volume drained of solute is proportional to $t^{3/2}$, and Eq. (1-36) results. If the precipitate particles were very long rods of fixed length, their volume or the volume of the region drained would change as r^2, and the last term in Eq. (1-36) would be replaced by a term in $(\sqrt{t})^2$ or t. Finally, if the precipitate formed in sheets, e.g., all over the grain boundary, the volume of the drained region would increase as $\sqrt{t}$, and the equation for $\bar{c}$ would change to the form $\bar{c} = c_o - \sqrt{\gamma t}$. Thus our model predicts that the precipitate shape can be determined from the initial slope of the plot of $ln(\bar{c} - c_o)$ versus t.

In the application of this analysis to experimental data, Eq. (1-36) is often replaces by an equivalent exponential equation. The function $c \exp(-bx)$ can be expanded into the series

$$c \exp(-bx) = c[1 - bx + (bx)^2/2! \ \ldots] \qquad (1\text{-}37)$$

which converges for all $bx < 1$. If $bx \ll 1$, the first terms $[c(1 - bx)]$ give a good approximation. Comparing Eqs. (1-36) and (1-37), it is seen that for short times we can write

$$\bar{c} = c_o[1 - (2bt/3c_o^{2/3})^{3/2}] = c_o \exp[-(t/\tau)^{3/2}] \qquad (1\text{-}38)$$

where the relaxation time τ is given by the expression

$$\tau = \frac{c''^{1/3} r_e^2 c_o^{2/3}}{2D(c_o - c')} = \frac{r_e^2}{2D} \left(\frac{c''}{c_o}\right)^{1/3} \qquad (1\text{-}39)$$

Ham shows that for $\alpha(0) = 0$, this exponential equation fits the data down to smaller values of $\bar{c}/c_o$ than does Eq. (1-34). This agreement is because of compensating errors.

The relaxation time τ can be determined from data on $\bar{c}(t)$. Usually the quantities c'', c' c_o, and D are known from other experiments. Thus r_e can be determined, and from this the mean interparticle spacing can be calculated. The quantities that contribute to τ, and thus vary the rate of precipitation can be noted in Eq. (1-39).

1.5   STRESS-ASSISTED DIFFUSION

General Effect of Potential Gradient. As the last problem to be solved for constant D, we consider the effect of an elastic stress gradient on diffusion. This is representative of the problems in which Fick's first law is no longer the flux equation. A potential gradient tends to produce a flux of atoms, and this flux must be added to that produced by the concentration gradient to arrive at the equation for the total flux. In this section we consider the effect of a general potential gradient on the flux equation and the resulting changes in the equation for $\partial c/\partial t$.

Consider a single particle moving in a potential field $V(x,y,z)$; the gradient of this potential describes the force F on the particle. Thus

$$\mathbf{F} = -\nabla V \tag{1-40}$$

As an example of this for macroscopic particles, consider a marble on an inclined plane. The potential here is due to gravity, and from elementary physics we know the force on the particle parallel to the plane to be proportional to the slope of the plane relative to the horizontal. Another common example of this type of force is the "pull" on a charged particle in an electrostatic potential gradient.

It is found empirically that a potential gradient of force gives rise to a mean diffusion velocity for the affected atoms. This fact is mathematically expressed by the equation

$$\mathbf{v} = M\mathbf{F} \tag{1-41}$$

in which M is called the mobility and has the units velocity per unit force. It is worthy of note that this equation is not of the form "force equals mass times acceleration." The force gives rise to a steady-state velocity instead of a continuing acceleration because on the atomic scale atoms are continually changing their direction of motion and thus cannot accelerate under the action of a force in the way a free particle does. This intermittent motion of the atoms on an atomic scale will be discussed in detail in the next chapter. It is purposely avoided in this entire chapter to give the reader a clearer picture of the type of problems that can be treated with no assumptions concerning the atomic processes involved.

In applying a potential gradient instead of a concentration gradient, we are simply replacing one small force with another. Thus it is plausible, or even necessary, that the mobility is proportional to the diffusion coefficient D. In Sec. 4-3, we show that the relationship is

$$M = D/kT \tag{1-42}$$

where k is Boltzmann's constant and T is the temperature in degrees Kelvin. The flux that results in a homogeneous system from F is thus the average velocity per particle times the number of particles per unit volume. If the units on J and c are consistent, we have

$$\mathbf{J} = c\mathbf{v} = M\mathbf{F}c = -(Dc/kT)\nabla V \tag{1-43}$$

If a concentration gradient exists in addition to ∇V, the flux equation is given by the addition of Eqs. (1-1) and (1-43) or

$$\mathbf{J} = -D(\nabla c + c\nabla V/kT) \tag{1-44}$$

Putting this flux equation in the continuity equation (1-7) gives, for constant D

$$\frac{\partial c}{\partial t} = D\nabla\left[\nabla c + \frac{c\nabla V}{kT}\right] \tag{1-45}$$

This then is the equation that needs to be solved to determine $c(x,y,z,t)$ in the presence of a potential gradient.

Solution for Very Short Times. The stress field around an interstitial atom in a solid solution is such that the atom can be attracted to a dislocation. Thus, in a supersaturated alloy, the precipitation rate on dislocations will be increased owing to the stress-induced drift which is superimposed on the drift due to any concentration gradient. If r is the radial distance between an interstitial atom and the core of an edge or screw dislocation, the interaction between the two can be approximated by the equations

$$V(r,\theta) = -B/r \quad \text{(screw)} \tag{1-46}$$

$$V(r,\theta) = -(A/r)\sin\theta \quad \text{(edge)} \tag{1-47}$$

where B and A are appropriately chosen constants.

If an alloy is homogenized at a high temperature and quenched to a low temperature where it is supersaturated, the initial distribution for an isolated dislocation is given by

$$c = c_o \quad \text{for} \quad r > 0, \quad \text{at} \quad t = 0$$

Expansion of Eq. (1-45) gives

$$\frac{\partial c}{\partial t} = D\nabla^2 c + \frac{D\nabla c\nabla V}{kT} + \frac{Dc\nabla^2 V}{kT} \tag{1-48}$$

Our aim is to determine the initial flux of atoms toward an isolated dislocation. This will depend on ∇c and ∇V. ∇V does not change with time, but ∇c does; and the general determination of how ∇c changes with time requires a solution of Eq. (1-48). However, for very short times the solution of this difficult problem can be avoided.

In the homogeneous alloy at $t = 0$, $\nabla c = 0$, so the last term in Eq. (1-48) determines $\partial c/\partial t$. The equation for $\nabla^2 V(r,\theta)$ in cylindrical coordinates is

$$\nabla^2 V(r,\theta) = \frac{\partial V^2}{\partial r^2} + \frac{1}{r}\frac{\partial V}{\partial r} + \frac{1}{r^2}\frac{\partial^2 V}{\partial\theta^2}$$

If $V(r,\theta)$ is given by Eq. (1-47), $\nabla^2 V = 0$, and $\partial c/\partial t = 0$. If $V = -\beta/$

r, then $\nabla^2 V \neq 0$; but the resulting change in concentration with time sets up concentration gradients slowly, and for short times the drift of solute toward the dislocation can be satisfactorily approximated by assuming that $\nabla c = 0.d$:[11] Using Eq. (1-46) because of its relative simplicity, we obtain $\nabla V = \beta/r^2$. Using Eq. (1-41) and (1-42), it is seen that this gradient moves the solute atoms toward the dislocation with a velocity given by the equation

$$v(r) = \frac{-dr}{dt} = \frac{D}{kT}\frac{B}{r^2} \qquad (1\text{-}49)$$

Integrating between $r = r'$ at $t = 0$ and $r = 0$ at $t = t'$ gives

$$r' = (3DBt'/kT)^{1/3} \qquad (1\text{-}50)$$

The interpretation of this equation is as follows. The atoms which were initially a distance r' from the core of the dislocation arrive at the dislocation core at $t = t'$, and other solute atoms which were initially at $r > r'$ have taken their places at $r = r'$. Thus, even though $\partial c/\partial t$ remains equal to zero, at $t = t'$ all solute which was in the region $r < r'$ at $t = 0$ will have precipitated or segregated at the dislocation core.

The amount of solute q removed per unit length of dislocation after t' is given by the expression

$$q = c_o \pi r'^2 = c_o \pi (3DBt'/kT)^{2/3} \qquad (1\text{-}51)$$

The period over which this solution is valid is determined by the volume in which the potential exerts an "appreciable" effect on the solute atoms. To be more precise, the thermal energy of a solute atom in the lattice will be about equal to kT. Thus when r becomes so large that $-V(r) < kT$, the potential energy will be less than the thermal energy of the particle, and the effect of the potential will be "inappreciable." We can thus define an "effective radius" for the potential as $r = R$, where

$$-V(R) = kT = B/R \qquad (1\text{-}52)$$

Thus the condition $\partial c/\partial t \simeq 0$ will hold longer in the region $r < R$, where ∇V has an effect, than in the region $r > R$, where the effect of ∇V is insignificant. The solution embodied in Eq. (1-51) can apply only for the solute initially in the region $r' < R$. For times when $r' > R$, appreciable concentration gradients are set up, and a different anal-

---

[11]A more rigorous and more complete discussion of the mathematical analysis of stress-assisted precipitation has been given by F. S. Ham, *J. Appl. Phys.*, *30* (1959) 915. A discussion of the approximations made herein is given there.

ysis of the problem is required. The value of R for the case of carbon in α-Fe can be estimated from a value of B. Taking $B \simeq 10^{-20}$ dyne-cm^2 gives $R \simeq 25$A.[12] With a dislocation density of $10^{11}/$cm^2, the mean distance between dislocations is about 300 A. Even with this relatively high dislocation density, Eq. (1-51) breaks down after only a small percentage of the solute has precipitated.

1.6 SOLUTIONS FOR VARIABLE D

All of the solutions discussed so far have been valid only for constant D. In real experiments the diffusion coefficient can, and will, vary. The diffusion coefficient for a given composition can vary with time, owing to changes in temperature. It can also change with composition, and since there is a concentration gradient, this means that D changes with position along the sample. In this latter case $D = D(x)$, and Fick's second law must be written

$$\frac{\partial c}{\partial t} = \frac{\partial}{\partial x}\left(D\,\frac{\partial c}{\partial x}\right) = \frac{\partial D}{\partial x}\frac{\partial c}{\partial x} + D\,\frac{\partial^2 c}{\partial x^2} \qquad (1\text{-}53)$$

The term $\partial D/\partial x$ makes the equation inhomogeneous, and the solution in closed form then ranges from difficult (for special cases) to impossible.

We will first discuss the solution for $D = D(c)$ which is most frequently used in solids and then show how to treat the case in which $D = D(t)$. For a more complete discussion of problems in which $D = D(c)$, see Crank.

Boltzmann-Matano Analysis. This is the solution for $D = D(c)$ most commonly referred to in metallurgical diffusion studies. It will serve as an example of the different line of attack required. It does not give a solution $c(x,t)$ as obtained before, but allows $D(c)$ to be calculated from an experimental $c(x)$ plot. If the initial conditions can be described in terms of the one variable $u = x/\sqrt{t}$, c is a function only of u, and Eq. (1-53) can be transformed into an ordinary homogeneous differential equation. Using the definition of u, we have

$$\frac{\partial c}{\partial t} = \frac{dc}{du}\frac{\partial u}{\partial t} = -\frac{1}{2}\frac{x}{t^{3/2}}\frac{dc}{du}$$

and

$$\frac{\partial c}{\partial x} = \frac{dc}{du}\frac{\partial u}{\partial x} = \frac{1}{t^{1/2}}\frac{dc}{du}$$

---

[12]A. Cochardt, G. Schoeck, H. Wiedersich, *Acta Met.*, *3* (1955) 533.

Substituting in the first part of Eq. (1-54), we obtain

$$-\frac{x}{2t^{3/2}}\frac{dc}{du} = \frac{\partial}{\partial x}\left(\frac{D}{\sqrt{t}}\frac{dc}{du}\right) = \frac{1}{t}\frac{d}{du}\left(D\frac{dc}{du}\right) \qquad (1\text{-}54)$$

or

$$\frac{-u}{2}\frac{dc}{du} = \frac{d}{du}\left(D\frac{dc}{du}\right) \qquad (1\text{-}55)$$

This transformation of Eq. (1-53) into Eq. (1-55) is due to Boltzmann. The method was first used to determine $D(c)$ by Matano.[13]

Consider the infinite diffusion couple which is described by the following initial conditions:

$$c = c_o \quad \text{for} \quad x < 0, \quad \text{at} \quad t = 0$$
$$c = 0 \quad \text{for} \quad x > 0, \quad \text{at} \quad t = 0$$

Since $x = 0$ is excluded at $t = 0$ and the original concentration is not a function of distance aside from the discontinuity at $x = 0$, the initial conditions can be expressed in terms of u only as

$$c = c_o \quad \text{at} \quad u = -\infty$$
$$c = 0 \quad \text{at} \quad u = \infty$$

Since Eq. (1-55) contains only total differentials, we can "cancel" $1/du$ from each side and integrate between $c = 0$ and $c = c'$, where c' is any concentration $0 < c' < c_o$

$$\frac{-1}{2}\int_{c=o}^{c=c'} u\,dc = \left[D\frac{dc}{du}\right]_{c=o}^{c=c'} \qquad (1\text{-}56)$$

The data on $c(x)$ are always at some fixed time so that substituting for u gives

$$\frac{-1}{2}\int_{o}^{c'} x\,dc = Dt\left[\frac{dc}{dx}\right]_{c=o}^{c=c'} = Dt\left(\frac{dc}{dx}\right)_{c=c'} \qquad (1\text{-}57)$$

The last equality in Eq. (1-56) comes from the fact that in this infinite system $dc/dx = 0$ at $c = 0$. From the additional fact that $dc/dx = 0$ at $c = c_o$, we have the condition

$$\int_{o}^{c_o} x\,dc = 0$$

so that Eq. (1-57) defines the plane at which $x = 0$. With this definition

---

[13]C. Matano, *Japan. Phys.*, **8** (1933) 109.

of x, $D(c')$ can be obtained from the graphical integration and differentiation of $c(x)$ using the equation

$$D(c') = \frac{-1}{2t}\left(\frac{dx}{dc}\right)_{c'}\int_o^{c'} x\,dc \qquad (1\text{-}58)$$

The quantities needed to calculate a value of D are shown in Fig. 1-7. The Matano interface is the plane at which $x = 0$ in Eq. (1-57). Graphically, it is the line that makes the two hatched areas of Fig. 1-7 equal. The value of D at $c = 0.2\,c_o$ would be calculated by measuring the cross-hatched area of the figure and the reciprocal of the slope. The error in the calculated value of $D(c)$ is largest at the ends of the curve where c/c_o approaches one or zero, since in these regions the integral is small and dx/dc large. To minimize these errors, the original concentration vs. distance data can be smoothed by plotting them on probability paper (one on which the error function is a straight line), and using the equation of that line to obtain dx/dc as a function of position.

This solution is quite useful for inferring $D(c)$ over a range of compositions in alloys. With the wide availability of the microprobe and computers, obtaining data on $D(c)$ can be highly automated. Eq. (1-58) assumes that the atomic volume of the alloy is independent of

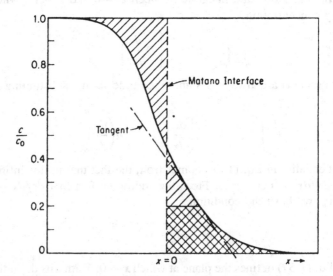

Fig. 1-7—The Matano interface is positioned to make the hatched areas on either side of it equal. The cross-hatched area and tangent show the quantities involved in calculating D at $c = 0.2c_o$.

composition, which is rarely true. For the use of a more generally valid equation see ref.[14]

Solutions for D as a Function of Time. If D is a function of time but not of position, inspection of Eq. (1-54) shows that the equation reduces to $\partial c/\partial t = D(t)(\partial^2 c/\partial x^2)$. What this means is that all of the solutions which were used for constant D can be used, but the product Dt must be replaced by an averaged product designated $\overline{Dt}$ and given by the equation

$$\overline{Dt} = \int_o^t D(t)\, dt \qquad (1\text{-}59)$$

The most common application of this equation is to correct for the diffusion that occurs during the heating and cooling of a diffusion couple which has been annealed at some fixed temperature, although it can also be used to calculate the degree of homogenization achieved during a complicated annealing cycle.

As an example of the application of Eq. (1-59), consider a diffusion couple that has the temperature-time history shown in Fig. 1-8a. The problem is to determine the time t' at temperature T' which would have produced the same amount of diffusion as actually occurred during the heating, annealing and cooling. This can be determined graphically once the T versus t data are transformed into a plot of D versus t. This has been done in Fig. 1-8b. It is seen that time spent in heating up to $0.8T'$ contributes nothing to the total amount of diffusion. This stems from the fact that $D(T)$ is given by an equation of the form

$$D = D_o \exp(-Q/RT)$$

For many cases Q is such that near the melting point of the metal D increases by a factor of 10 for each increase of 10% in the absolute temperature. Figure 1-8b was obtained by using this relationship.

1.7 TWO-PHASE BINARY SYSTEMS

Binary systems seldom exhibit complete solid miscibility. A more common situation is the simple eutectic system with limited solubility on each side of the diagram. If A-B alloys are described by the phase diagram in Fig. 1-9, what would the $c(x)$ curve be for a diffusion

---

[14]H. C. Akuezue, D. P. Whittle, *Metal Science*, *17* (1983) 27. For a FORTRAN program for solving the Matano-Boltzmann equation see "MATAN0—A Computer Code for the Analysis of Interdiffusion and Intrinsic Diffusion Information in Binary Systems," P. T. Carlson, ORNL-5045, June 1975, NTIS, 5285 Port Royal Rd., Springfield VA, 22161.

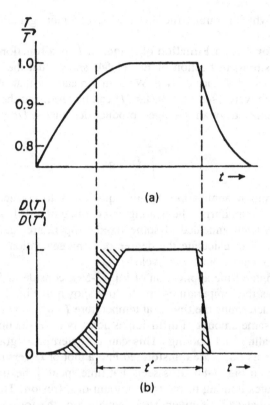

Fig. 1-8 — (a) Temperature versus time record of diffusion sample; (b) D vs. time for same sample. t' is the time at T' which would give the same amount of diffusion as actually occurred in the cycle.

couple made by joining pure A to pure B and annealing at a temperature below the eutectic temperature?

If the equilibrium solubilities are maintained in the two phases at the alpha/beta interface, Fig. 1-9 shows the $c(x)$ curve that develops during a diffusion anneal. Note how the concentration of B drops from the solubility limit of alpha in beta, $C_{\beta\alpha}$, to that for beta in alpha, $C_{\alpha\beta}$. That is, no two-phase $\alpha + \beta$ region is generated in the diffusion couple. This is due to the fact that no concentration gradient can exist across a 2-phase field in a binary system. Another way to look at this is to consider the drop in the chemical potential of B, μ, across the diffusion couple. This is also shown in Fig. 1-9. The chemical potential is continuous across the interface, though there may be a change in slope.

Another measurable and useful quantity in this type of diffusion couple is the rate at which the two phase interface moves under the in-

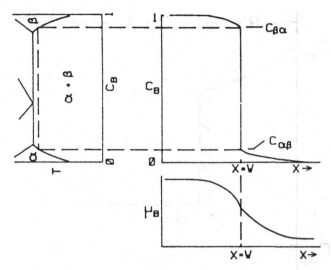

Fig. 1-9—c(x) for diffusion couple between A and B which exhibit a simple eutectic phase diagram. The variation of the chemical potential of B across the couple is also shown.

fluence of diffusion. It is usual to assume that equilibrium exists across the $\alpha - \beta$ interface. Thus the compositions of the two phases at the interface are constant and given by the phase diagram. Under these conditions one equation can be obtained from the conservation of material at the advancing interface. Considering the diffusion couple presented in Fig. 1-9, the rate of advance of the interface between the α and β is proportional to the difference in the flux into the interface ($J_{\alpha\beta}$) and the flux out ($J_{\beta\alpha}$)

$$[C_{\alpha\beta} - C_{\beta\alpha}](dw/dt) = J_{\alpha\beta} - J_{\beta\alpha}$$

$$= \left[-D_\alpha \left(\frac{\partial c}{\partial x}\right)_{\alpha\beta} + D_\beta \left(\frac{\partial c}{\partial x}\right)_{\beta\alpha} \right] \quad (1\text{-}60)$$

Constant D. If the diffusion coefficients in Eq. (1-60) are independent of composition, there are straightforward means for treating the problem for different boundary conditions. As an example consider a system in which solute is added to the surface of a two phase alloy of composition C_o. As B is added to the surface it is all used to convert α to β. There will be no flux of B out of the surface layer of beta into the two phase alloy. Fig. 1-10 shows $c(x)$ for this system. If the surface composition is held at a value C_s near pure B, a layer of β develops on the surface whose thickness is designated as w. The boundary conditions for $c(x)$ in the beta are

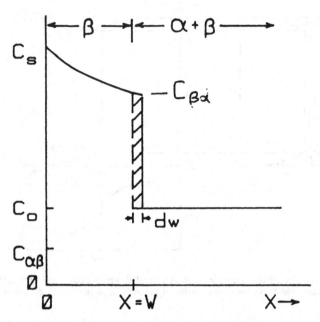

Fig. 1-10—$c(x)$ for the diffusion controlled growth of a B rich beta layer into an initially homogeneous two phase alloy. The hashed area indicates amount of solute needed to advance the beta phase by dw.

$$
\begin{aligned}
c &= C_s \quad \text{at} \quad x = 0 \quad \text{and} \quad t > 0 \\
c &= C_{\beta\alpha} \quad \text{at} \quad x = w \quad \text{and} \quad t > 0
\end{aligned} \tag{1-61}
$$

With these Eq. (1-60) becomes

$$
[C_{\beta\alpha} - C_o](dw/dt) = J_{\beta\alpha} = -D\left(\frac{\partial c}{\partial x}\right)_{\beta\alpha} \tag{1-62}
$$

The simplest approach is to approximate the gradient in the beta phase by a straight line of slope $-(C_s - C_{\beta\alpha})/w$. Eqn. 1-62 would then integrate to

$$
w^2 = [(C_s - C_{\beta\alpha})/2(C_{\beta\alpha} - C_o)]4Dt \tag{1-63}
$$

A better approximation is obtained if $c(x,t)$ in the beta layer is described by an error function solution to the diffusion equations. This would be exact for a fixed interface, and since the interface is moving slowly it is a good approximation. The solution for $c(x,t)$ in the beta phase is then

$$
c(x,t) = C_s - H \, \mathrm{erf}(x/2\sqrt{Dt}), \quad \text{for} \quad 0 < x < w \tag{1-64}
$$

H is a constant involving concentrations. The width of the beta layer should be given by an equation of the form

$$w^2 = \gamma^2 \, 4Dt \qquad (1\text{-}65)$$

where γ is a constant to be determined. Thus at $x = w$

$$C_s - C_{\beta\alpha} = H \, \text{erf}(\gamma) \qquad (1\text{-}66)$$

The derivative of Eq. (1-65) give an equation for the gradient in the beta phase as x approaches w. Eq. (1-62) and (1-64) then give[15]

$$C_{\beta\alpha} - C_o = H/(\gamma\sqrt{\pi}) \exp(-\gamma^2)$$

Eliminating H between these last two equations

$$(C_s - C_{\beta\alpha})/(C_{\beta\alpha} - C_o) = \gamma\sqrt{\pi} \exp(\gamma^2) \, \text{erf}(\gamma) \qquad (1\text{-}67)$$

This equation is valid for a wide range of values of γ. If $(C_s - C_{\beta\alpha}) \ll (C_{\beta\alpha} - C_o)$ then γ is small. In such cases $\exp(\gamma^2) \simeq 1$ and $\text{erf}(\gamma) \simeq 2\gamma/\sqrt{\pi}$. Or, substituting into Eq. 1-64

$$w^2 = [(C_s - C_{\beta\alpha})/2(C_{\beta\alpha} - C_o)] \, 4Dt \qquad (1\text{-}68)$$

which is identical to the result of the linear approximation, Eq. (1-63). However, for γ larger than 0.1 its value must be obtained from Eq. (1-67).

Using Eq. (1-64), γ can be obtained for any ratio of concentrations, and the value of D in the growing phase determined from measurements of the rate of growth of that phase. This same procedure can be extended to the case of a diffusion couple made by joining pure A to pure B with concentration gradients on both sides of the advancing interface.[16]

Variable D. When D changes significantly with composition there is no closed solution to the diffusion equation. However, returning to the diffusion couple shown in Fig. 1-9, note that this system fulfills the conditions required for the application of the Matano-Boltzmann solution, namely the initial conditions can be expressed in terms of the function $u = x/\sqrt{t}$. The Matano-Boltzmann formulation can be used to obtain some simple, useful equations. For example the thickness of intermetallic phases (in binary systems) increase as $\sqrt{t}$.[17] The relation between the rate constant for diffusion couples developing only one phase and those developing several has been developed by Shatynski et al.[18]

---

[15] See Problem 1-3 for the differentiation of an error function.
[16] W. Jost, *Diffusion in Solids, Liquids, Gases*, Academic Press, 1952, pp. 69–75.
[17] G. V. Kidson, *J. Nucl. Matl.*, *3* (1961) 21.
[18] S. R. Shatynski, J. P. Hirth, R. A. Rapp, *Acta Met.*, *24* (1975) 1071.

1.8   DIFFUSION IN NONCUBIC LATTICES

Up to this point we have assumed that $D(=J/\nabla c)$ is a constant which is independent of the direction of the flux relative to the crystallographic axes. In this section it will be shown that this assumption is true in any crystalline solid whose lattice has cubic symmetry. However, in noncubic crystals a complete description of the relation between the flux J and the concentration gradient ∇c requires the knowledge of two or more constants. As with the other topics discussed in this chapter, the proof of these properties of D requires no assumption about the atomic processes involved in diffusion. Rather, the results follow from the symmetry of the crystal and the properties of a second-order tensor.

The aim of the treatment given here is to develop an understanding of the physical basis of the results. The reasoning is rigorous but lacks elegance and generality. A general treatment requires a knowledge of the transformation properties of second-order tensors. General treatments can be found in Wooster and in Nye.[19]

Consider the two vectors J and ∇c. In the most general case they will not be parallel, so that the simple relation $J = D\nabla c$, where D is a constant, is not adequate. For the general case it is assumed that a given component of the flux is influenced by each of the components of the gradient. Thus, the equations are[20]

$$J_x = -D_{11}\frac{dc}{dx} - D_{12}\frac{dc}{dy} - D_{13}\frac{dc}{dz} \qquad (1\text{-}69)$$

$$J_y = -D_{21}\frac{dc}{dx} - D_{22}\frac{dc}{dy} - D_{23}\frac{dc}{dz} \qquad (1\text{-}70)$$

$$J_z = -D_{31}\frac{dc}{dx} - D_{32}\frac{dc}{dy} - D_{33}\frac{dc}{dz} \qquad (1\text{-}71)$$

where the various scalar fluxes are parallel to the three axes of a cartesian coordinate system. The set of nine numbers designated D_{ij} is called a second-order tensor and is defined by the above equations. To demonstrate the possible simplification of these equations, consider a cube of material whose lattice has cubic symmetry. Imagine that this cube

---

[19] W. Wooster, *A Text Book on Crystal Physics*, Chap. 1, Cambridge Univ. Press, New York, 1949. A more general treatment is given by J. Nye, "Physical Properties of Crystals," Chap. 1, Clarendon Press, 1957.

[20] A moment's reflection will show that these equations, and thus the treatment given here, can easily be adapted to the case of heat flow, the flow of electricity, and other cases in which a vector flux is related to a potential gradient. Thus, the results obtained below for D_{ij}, hold equally well for the thermal and electrical conductivities.

is a single crystal with the cubic axes of the lattice perpendicular to the cube faces. A cartesian coordinate system is now set up with axes perpendicular to the cube faces (parallel to the cube axes of the lattice), and a concentration gradient is established such that $dc/dx = \alpha \neq 0$ and $dc/dy = dc/dz = 0$. Equation (1-69) then gives the component of the flux parallel to the gradient as

$$J_x = -D_{11}(dc/dx)$$

If this gradient is removed and replaced by an equal gradient along the y axis, that is, $dc/dy = \alpha$ and $dc/dx = dc/dz = 0$, the component of the flux parallel to the gradient will be

$$J_y = -D_{22}(dc/dy)$$

Now if the lattice of the specimen has cubic symmetry, the x axis ([100] direction) and the y axis ([010] direction) are indistinguishable. Since the gradients were of equal magnitude in both cases this symmetry requires that J_x in the first experiment must equal J_y in the second. But this says that $D_{11} \alpha = D_{22} \alpha$, which requires that $D_{11} = D_{22}$. Similarly the z axis ([001] direction) is indistinguishable from the x axis, so for the cube crystal

$$D_{11} = D_{22} = D_{33} \tag{1-72}$$

In a lattice with tetragonal symmetry, the [010] and the [100] directions are identical but are distinguishable from the [001] direction. Thus in a tetragonal lattice

$$D_{11} = D_{22} \neq D_{33} \tag{1-73}$$

Finally, for an orthorhombic lattice each of the [100], [010], and [001] directions are distinguishable, so

$$D_{11} \neq D_{22} \neq D_{33}$$

For the case of hexagonal crystals, the above type of argument will prove that D_{11} is the same in each of the six close-packed directions of the basal plane. However, the fact that $D_{11} = D_{22}$, for the case of orthogonal axes in the basal plane cannot be proved without recourse to the transformation properties of second-order tensors. If it is accepted that $D_{11} = D_{22}$, then it should be apparent that of hexagonal crystals

$$D_{11} = D_{22} \neq D_{33}$$

To enlarge on these results it is possible that in some hexagonal lattice D_{22} may be found experimentally to equal D_{33}. However, this cannot be asserted a priori from the symmetry. On the other hand if it is reported that $D_{11} \neq D_{22}$ for some hexagonal material, either the experimental results are wrong or some type of defect, or field, was present which, in effect, destroyed the sixfold rotational symmetry about the [001] axis. The only other alternative would be to conclude that Eqs. (1-69) to (1-71) are inapplicable. This is conceivable but extremely improbable.

So far we have ignored the constants D_{ij} where $i \neq j$. However, if D_{ij} is to be a simple constant for any cubic crystal, then in addition to proving that $D_{11} = D_{22} = D_{33}$, we must show that all $D_{ij} = 0$, if $i \neq j$. This can be done as follows. Again consider the single crystal with a cubic lattice. We establish a gradient such that $dc/dy = \alpha$, but $dc/dx = dc/dz = 0$. By measuring the component of the flux along the x axis, we can determine D_{12} since Eq. (1-69) gives

$$J_x = -D_{12}(dc/dy) = -D_{12}\alpha \qquad (1\text{-}74)$$

Imagine next that the source and the sink which established this gradient are removed and the crystal is rotated 180 degrees about its x axis ([100] direction), relative to the source and sink. Since the [100] direction has fourfold rotational symmetry, the original and final positions of the lattice will be indistinguishable. If the source and sink are again applied, the gradient is again established along the y axis. However, since the y axis is fixed in the crystal, the 180 deg. rotation has interchanged the $+y$ and the $-y$ axes, so the gradient will be just the negative of its previous value. This means that now $dc/dy = -\alpha$, and $dc/dx = dc/dz = 0$. This gives

$$J_x = -D_{12}(dc/dy) = D_{12}\alpha \qquad (1\text{-}75)$$

But by symmetry, J_x from Eq. (1-74) must equal J_x from Eq. (1-75), or

$$D_{12} = -D_{12}$$

The required symmetry can be satisfied only if $D_{12} = 0$. If the gradient had been along the z axis, the same rotation could have been used to show that $D_{13} = 0$. Since the y and z axes also have twofold rotational symmetry, it follows that all of the off-diagonal terms (that is, D_{ij} with $i \neq j$) are zero. Finally we have used only twofold rotational symmetry, so from the same proof it follows that the D_{ij} also equal zero when $i \neq j$ for tetragonal and orthorhombic crystals.

Similar reasoning shows that the $D_{ij}(i \neq j)$ terms are zero for lattices which have hexagonal symmetry. The results can be summarized as follows:

$$\text{Cubic} \qquad D_{ij} = \begin{bmatrix} D_{11} & 0 & 0 \\ 0 & D_{11} & 0 \\ 0 & 0 & D_{11} \end{bmatrix} \qquad (1\text{-}76)$$

$$\text{Textragonal} \atop \text{\& hexagonal} \quad D_{ij} = \begin{bmatrix} D_{11} & 0 & 0 \\ 0 & D_{11} & 0 \\ 0 & 0 & D_{33} \end{bmatrix} \qquad (1\text{-}77)$$

$$\text{Orthorhombic } D_{ij} = \begin{bmatrix} D_{11} & 0 & 0 \\ 0 & D_{22} & 0 \\ 0 & 0 & D_{33} \end{bmatrix} \qquad (1\text{-}78)$$

To demonstrate the use of these conclusions and Eqs. (1-69,71), we shall apply them to the case where D_{ij} is given by Eq. (1-77). Consider the determination of the diffusion coefficient in such a material using a tracer. If a single crystal is available with the c axis (the sixfold axis in hexagonal or fourfold in tetragonal) normal to one face, the tracer can be deposited on that face. After diffusion the concentration gradient is parallel to the c axis, since $D_{13} = D_{23} = 0$, and so is the flux. If the concentration distribution $c(u)$ is determined by taking sections parallel to the initial face, one obtains D_{33} from a plot of lnc versus u^2. If the c axis is in the face covered with the tracer, the flux and the gradient are again parallel, and the determination of D_{11} is straightforward.

For the intermediate case where the plated face makes an angle of $90 - \theta$ with the c axis, the concentration gradient makes an angle of θ with the c axis. In this case the flux and the gradient are not parallel if $D_{11} \neq D_{33}$. This is shown in Fig. 1-11, where $D_{11} < D_{33}$. If $c(u)$ is

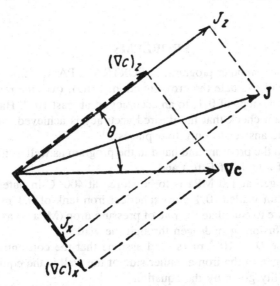

Fig. 1-11—The orientation of the flux $\mathbf{J}$ and the concentration gradient ∇c for the case of $D_{11} < D_{33}$. The axis is parallel to the c axis of the lattice. ∇c is normal to the y axis so that $(\nabla c)_y$ and J_y equal zero.

Table 1-2. $D\perp$ and $D\|$ for Several Metals

Metal	Struc	$D_o\|$ cm²/s	$D_o\perp$ cm²/s	$Q\|$ kJ/mol	$Q\perp$ kJ/mol	$D\perp/D\|$ $T = 0.8T_m$
Be	hcp	0.52	0.68	157	171	0.31
Cd	hcp	0.18	0.12	82.0	78.1	1.8
α-Hf	hcp	0.28	0.86	349	370	0.87
Mg	hcp	1.5	1.0	136	135	0.78
Tl	hcp	0.4	0.4	95.5	95.8	0.92
Sb	hex	0.1	56	149	201	0.098
Sn	bct	10.7	7.7	105	107	0.40
Zn	hcp	0.18	0.13	96.4	91.6	2.05

Data from N. L. Peterson, *Jnl. Nucl. Matl.*, *69&70* (1979) 3–37.

now determined, the value of D obtained from a plot of lnc versus u^2 is called $D(\theta)$ and is given by the equation

$$D(\theta) = D_{33} \cos^2 \theta + D_{11} \sin^2 \theta \qquad (1\text{-}79)$$

If necessary D_{33} and D_{11} can be determined from measurements of $D(\theta)$ for different values of θ.

Data for some noncubic materials is given in Table 1-2. Here $D\perp$ and $D\|$ refer to the diffusion coefficients perpendicular and parallel to the c axis of the lattice.

PROBLEMS

1-1. Write a computer program, in FORTRAN, PASCAL, or BASIC, that will evaluate the error function, Erf(z), over the range $0 < z < 2$ in steps of 0.1, to an accuracy of at least 10^{-4}. Have your program check that the desired accuracy is achieved, and print out the answer to only four places.

Run the program and hand in the program as well as the printout of a table of Erf(z) and z.

1-2. Hydrogen at 1.0 MPa is to be stored at 400° C in outer space, in a thin walled (0.1 mm) spherical iron tank of 0.1 m radius. You are to calculate the rate of pressure drop (MPa/s) as a result of diffusion of hydrogen through the wall.

Take $D = 10^{-8}$ m²/s, and assume that the concentration of hydrogen in the iron at either side of the wall is the equilibrium solubility given by the equation

$$C(P) = 10^{-5} (P)^{1/2} \text{ gmH/gmFe} \quad (P \text{ in MPa}).$$

1-3. By differentiation show that $\text{Erf}(x/2\sqrt{Dt})$ is a solution to Eq. (1-13). Note

$$\text{If } F(u) = \int_{a(u)}^{b(u)} f(x)dx, \quad \text{then}$$

$$\frac{dF}{du} = -f(a(u)) \frac{da}{du} + f(b(u)) \frac{db}{du}$$

1-4. A thin film of radioactive copper was electroplated on the end of a copper cylinder. After a high temperature anneal of 20 hr, the specimen was sectioned, and the activity of each section counted. The data are:

a (counts/s/mg)	x (0.01 cm)
5012	1
3981	2
2512	3
1413	4
525	5

(a) Plot the data and determine D from the best line plotted by eye.

(b) Calculate the slope of log a versus x^2 using a least-square procedure, and plot the least squares line on the figure for part (a).

1-5. A piece of 0.1% C steel is to be carburized at 930° C until the carbon content is raised to 0.45% C at a depth of 0.05 cm. The carburizing gas holds the surface at 1% carbon for all $t > 0$. If $D = 1.4 \times 10^{-7}$ cm^2/s for all compositions,

(a) Calculate the time required at the carburizing temperature.

(b) What time is required at the same temperature to double this amount of penetration?

(c) If $D = 0.27 \exp(-17,400/T)$ cm^2/s, what temperature increase is required to get 0.45% C at a depth of 0.1 cm in the same time as 0.05 cm was attained at 930° C?

1-6. If helium is injected into a thin (100 nm) copper foil and then the foil is heated, the He quickly combines with vacancies to form very small bubbles (10 nm radius). These bubbles diffuse about in a random manner just as atoms do. However, when they touch a free surface, i.e. come within a radius of the surface, they are annihilated. When observing the foil in the TEM, between consecutive anneals, it is much easier to measure the number of bubbles per unit area than it is to measure the distribution of bubbles through the thickness.

(a) For an initially uniform concentration of bubbles write the boundary conditions and give the equation for the average concentration through the foil as a function of time, $\bar{c}(t)$, for a foil of thickness h and bubble diffusion coefficient, D_b.

(b) Often the original distribution of bubbles is not uniform through the thickness, but is higher on one side. Show how this would affect a plot of $ln\bar{c}(t)$ vs $t.$, and explain how you could still get D_b from the data.

1-7. It is suggested that the change in hydrogen potential of an aqueous solution could be measured by the change in electrical resistance of an inert metal foil immersed in the solution (The resistance of the foil is proportional to the average concentration of hydrogen in the metal.) If the diffusion coefficient of hydrogen in the metal is 10^{-5} cm^2/s, what should the thickness of the foil be if the concentration in the foil is to reach 0.95 of its final value in one second.

1-8. A thick walled steel pressure vessel in an oil refinery contains high pressure hydrogen for a long time. To avoid hydrogen cracking on cooling the vessel is to be held at temperature with no hydrogen inside it until most of the H has diffused out. As boundary conditions take for $c(x,t)$:

$$c(x,0) = c_o\, x/h \quad \text{at} \quad t = 0$$

$$c(0,t) = 0 = c(h,t) \quad \text{at} \quad t > 0.$$

(a) Derive $c(x,t)$ for $t > 0$.

(b) If $D = 5.8 \times 10^{-4}$ cm^2/s, and $h = 25$ cm, how long does it take to get the average concentration, $\bar{c}$, reduced to 0.1 c_o?

1-9. A sensitive infra-red detector is made by vapor depositing alternate 13 nm thick layers of HgTe and CdTe on a substrate. It is found that the detector loses its sensitivity due to interdiffusion after 55 hr at 162° C or after 250 hr at 110° C.

(a) Calculate an activation energy for the interdiffusion process and estimate the life of the device at 25° C.

(b) Estimate the interdiffusion coefficient at 162° C.

1-10. In a pure gold quenched from 700° C, it is thought that the supersaturation of vacancies is relieved by adsorption of vacancies at dislocation lines.

(a) Considering the dislocation lines to be fixed cylindrical sinks of constant radius r_o, derive an equation giving the time dependence of the ratio of the average vacancy concentration

$\bar{c}(t)$ to the initial concentration c_o (for $0.8 < \bar{c}/c_o < 1$), which could be used to check this hypothesis.

(b) Derive an equation for the case in which planar grain boundaries act as sinks for the vacancies.

1-11. If a disc of an initially homogeneous single phase alloy of Ag in Pb is spun in a centrifuge at a temperature where diffusion can occur the lighter Ag atoms will diffuse toward the top of the disc and the Pb toward the bottom. A sample is spun at 100,000 g until $c(x)$ attains a steady state. Using Eq. 1-46, answer the following questions:

(a) Give an equation relating the concentration gradient to the acceleration and density difference.

(b) Does the concentration gradient depend on D? What determines the value of $d \ln c/dx$?

(c) How would you guess the relaxation time for reaching this steady-state is related to the height of the disc and D?

1-12. An edge dislocation glides into a homogeneous region of an Fe-C solid solution of C concentration c_o and comes to rest:

(a) Integrate Eq. 1-45 to obtain an equation for $c(r)$, the distribution of carbon around the dislocation after "infinite" time.

(b) If the radius within which carbon redistributes in the elastic field of the dislocation is 1 nm, estimate the time to redistribute the carbon in this region, i.e. from above the glide plane to below it. (take D for C in ferrite to be 10^{-20} m^2/s, which is true for 80° C). Explain the assumptions made in obtaining your estimate.

1-13. Metals 'A', and 'B' form a simple eutectic diagram of the type shown in Fig. 1-9. A diffusion couple between pure A and pure B is made with a thin $\alpha + \beta$ layer in the middle, by means of powder metallurgical techniques.

(a) The couple is given a brief anneal. In one figure draw a $c(x)$ curve that extends from the pure A region of the diffusion couple to pure B. (Note there will be two $c(x)$ curves in the two phase region, one for each phase.)

(b) Consider the fluxes out of each side of the two-phase region and show that the two-phase region will shrink with time.

1-14. If pure A is joined to pure B (Fig. 1-9) there is a flux of A into the B-rich phase (β), and B into the A-rich phase (α). The difference between these two fluxes is reflected in a shift of the $\alpha - \beta$ interface relative to the end of the specimen. Starting with Eq. 1-60, use a procedure similar to that developed in the text to derive an equation relating the rate of this shift, dw/dt, to the diffusion coefficients and concentration gradients in α and β for the case in which D in both phases are independent of composition and thus error function solutions are valid. (For help see G.H. Geiger, D.R. Poirier, *Transport Phenomena in Metallurgy*, (Addison Wesley, 1975), pp. 490–6.

1-15. An Fe-0.8% C alloy is decarburized in an atmosphere that keeps the surface essentially carbon free.

   (a) If the decarburization is carried out at 730° C a carbon free layer of ferrite forms on the surface. Derive an equation relating the alpha layer to the concentration difference across the ferrite layer, time, and D in the the ferrite.

   (b) Assuming $C_{\alpha\gamma} = 0.02$ w/o carbon and that $D = 10^{-6}$ cm^2/s for carbon in the ferrite, how long will it take to form a ferrite layer 0.01 cm thick.

   (c) If the decarburization is carried out at 800° C the $c(x)$ curve crosses the alpha, alpha+aust. and aust. fields. Plot the $c(x)$ curve across the diffusion zone. On the plot label the composition limits of the two-phase fields.

1-16. The general equations for diffusion in a two-dimensional lattice are:

$$J_x = -D_{11}(dc/dx) - D_{12}(dc/dy)$$

$$J_y = -D_{21}(dc/dx) - D_{22}(dc/dy)$$

Show what elements of the diffusion tensor, D_{ij}, are zero and which are equal for:

   (a) A square array of points.

   (b) A rectangular array of points.

(This treatment should be valid for the diffusion of tracer atoms on the face of a crystal.)

1-17. For a hard brittle material it is difficult to determine D by grinding off layers and collecting the material removed. An alternative is to count the activity remaining in the sample after removing material by grinding.

   (a) For the case in which a thin layer of tracer is placed on the original surface at $x = 0$ give the solution to the diffusion equation, $c(x,t)$, after a diffusion anneal, but before any material is ground off the surface. Draw $c(x,t)$ vs. x.

   (b) Show that the total amount of material left in the sample after the diffusion layer has been ground away to a depth of d is:

$$q(d) = B[1 - \text{erf}(d/2\sqrt{Dt})]$$

For an isotope whose radiation is adsorbed little in leaving the sample, B is the total activity before any grinding is done. Note the determination of $q(d)$ allows the determination of D.

Answers to Selected Problems

1-2. Initial $dP/dt = 670$ Pa/s.

1-4. $D = 3.73 \times 10^{-9}$ cm²/s.

1.5. (a) 1.2×10^4 s.
(b) Quadruple the time of (a)
(c) 128° C

1-6. (a) For $t = 0$, $c(x,0) = c_o$; for $t > 0$, $c(r,t) = c(h - r,t) = 0$
$\bar{c}(t)/c_o = (8/\pi^2)(\exp(-t/\tau) + (1/9) \exp(-9t/\tau) + ...)$
(b) $c(x,t)$ now contains terms like $\sin(2\pi x/h)$ so $c(t)$ has terms in $\exp(-4t/\tau)$. Thus $ln\bar{c}$ vs t is not linear until $t > \tau/4$.

1-7. 0.006 cm.

1-8. (a) $c(x,t) = (2 \ c_o/\pi)[\sin(\pi x/h)\exp(-t/v) - (1/2)\sin(2\pi x/h)\exp(-4t/\tau) + (1/3)\sin(3\pi x/h)\exp(-9t/\tau) - ...]$
(b) $\tau = h^2/D\pi^2 = 1.1 \times 10^{+5}$ s. To remove 90% requires $t = -\tau \ ln(0.1 \ \pi^2/4)$.

1-9. (a) 40.3 kJ/mol, 9300 hr.
(b) 8.5×10^{-19} cm²/s.

1-11. If the atomic volume of Ag and Pb are assumed to be the same, and $m(Ag)$ is the mass of a silver atom, then
(a) $d \ lnc(x)/dx = 10^5 \ g[m(Pb) - m(Ag)]/kT$.

1-12. (a) $c = c_o \exp(-V/kT) = c_o \exp(-A \sin\theta/rkT)$
(b) Approximating this as relaxation in a cylinder, $\tau \approx r^2/D = 100$ s.

1-15. (a) $w^2 = [C_{\alpha\gamma}/(C_{\gamma\alpha} - C_{\alpha\gamma})] 2D_\alpha t$
(b) 1950 s.

1-16. (a) $D_{21} = D_{12} = 0$, $D_{11} = D_{22}$.
(b) $D_{21} = D_{12} = 0$, $D_{11} \neq D_{22}$.

2

ATOMIC THEORY OF DIFFUSION

If a drop of a dilute mixture of milk in water is placed under a microscope and observed by transmitted light, small fat globules can be seen. These globules are about 1 μm in diameter and continually make small movements hither and yon. These movements, which are called Brownian motion, give a continual mixing and are the cause, or mechanism, of the homogenization, whose rate could be measured in a macroscopic diffusion experiment. For example, if a drop of the same milky solution is placed in water, it will tend to spread out, and the mixture will ultimately become homogeneous. In this latter experiment a concentration gradient is present, a flux of fat globules[1] exists, and a diffusion coefficient could be measured. This is not quite an after-lunch experiment though, since turbulent mixing must be avoided and diffusion occurs quite slowly ($D = 10^{-8}$ cm^2/sec).

Brownian motion is not peculiar to the fat droplets in milk; in fact, active study at the turn of the century showed that it occurred for any microscopic particles suspended in any liquid or gas. This being the case, there is the interesting and potentially complicated question of how the random motion of these particles is related to the macroscopic displacement of the particles. For example, given the number of jumps per second and the mean jump distance, how far will the particle be from an arbitrary starting point after some very large number of jumps? This particular problem was initially treated about 1905 by Smolu-

---

[1]The composition of milk is not simple, but for our purpose it can be considered a colloidal dispersion of fat globules in water. Milk is used as an example because it is the most easily obtained dispersion with particles that can be resolved at 500X.

chowski and by Einstein and has been further developed over the years.[2] It is generally called the random-walk problem.

This may seem like a digression from diffusion in crystalline solids, but the problems have striking similarities. It is impossible to observe the motion of the individual atoms in solids, but diffusion does occur in them, so there must be relative motion of the atoms. It is therefore reasonable to assume that diffusion occurs by the periodic jumping of atoms from one lattice site to another. If this is indeed true, then the mathematics of the random-walk problem will allow us to go back and forth between the observed macroscopic diffusion coefficients and the jump frequencies and jump distances of the diffusing atoms. The problem is not a simple one, but it is most exciting since it transforms the study of diffusion from the question of how fast a system will homogenize into a tool for studying the atomic processes involved in a variety of reactions in solids, for studying defects in solids, or for studying the interaction between the atoms themselves.

In this chapter we discuss the random-walk problem, the atomic mechanisms which are thought to give rise to diffusion, the factors which influence the jump frequency of the atoms, and the calculation of a diffusion coefficient from the combination of all of these.

2.1 RANDOM MOVEMENT AND THE DIFFUSION COEFFICIENT

Before discussing the detailed mechanisms and mathematics of diffusion, it is helpful to study a simple situation in which no detailed mechanism is assumed. In this section we shall derive an approximate equation relating D to the jump frequency and the jump distance without going through a rigorous treatment of the random-walk problem.

Consider a crystalline bar that has a concentration gradient along its y axis (see Fig. 2-1). We consider only jumps to the left and right, that is, those giving a change in position along the y axis. Consider now two adjacent lattice planes, designated 1 and 2, a distance β apart. Let there be n_1 diffusing atoms per unit area in plane 1 and n_2 in plane 2. If each atom jumps an average of Γ times per second, the number of atoms in plane 1 that jump in the short period dt is $n_1 \Gamma dt$. Assuming that the jump frequency is the same in all orthogonal directions, one-

---

[2]A readable, interesting treatment of Brownian motion can be found in the translation of Einstein's original papers. (A. Einstein, *Investigations on the Theory of Brownian Movement*, Dover Publications, New York, 1956.) For an advanced, complete treatment which deals more with mathematics and less with physical phenomena, see N. Wax (ed.), *Selected Papers on Noise and Stochastic Processes*, Dover Publications, New York, 1954.

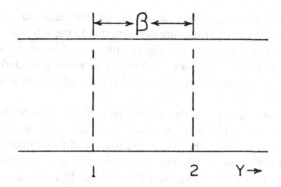

Fig. 2-1

sixth of the atoms will go to the right to plane 2, the number of atoms jumping from plane 1 to plane 2 in dt is $(1/6)n_1\Gamma dt$. The net flux from planes 1 to 2 is thus

$$J = \frac{1}{6}(n_1 - n_2)\Gamma = \frac{\text{number of atoms}}{\text{(area) (time)}}$$

The quantity $(n_1 - n_2)$ can be related to the concentration or number per unit volume by observing that $n_1/\beta = c_1$ and $n_2/\beta = c_2$, giving

$$J = (1/6)(c_1 - c_2)\beta\Gamma$$

But in essentially all diffusion studies, c changes slowly enough with position that

$$c_1 - c_2 = -\beta(\partial c/\partial y)$$

so that

$$J = -\frac{1}{6}\beta^2\Gamma\frac{\partial c}{\partial y} \qquad (2\text{-}1)$$

This equation is identical to Fick's first law if the diffusion coefficient D is given by

$$D = (1/6)\beta^2\Gamma \qquad (2\text{-}2)$$

The diffusion coefficient is therefore determined by the product of the jump distance squared and the jump frequency.

It should be emphasized that Γ was assumed to be the same for jumps from left to right as from right to left. Thus the flow down the concentration gradient does not result from any bias of the atoms to jump in that direction. If each atom jumps randomly and $n_1 > n_2$, there will be a net flux from 1 to 2 simply because there are more atoms on plane 1 to jump to 2 than there are atoms on 2 to jump to 1. (If the jump frequency to the right was greater than that to the left there would be a net drift of the atoms to the right. This sort of an effect will be discussed in subsequent chapters, for example when considering the case of an electric potential gradient on diffusion of ions in ionic solids, in Chap. 5.)

Without assuming a particular mechanism, it is plausible that β is about the interatomic distance in a lattice, or the order of one Angstrom. If we assume this, the jump frequency can be estimated from the measured diffusion coefficient. For carbon in β-Fe at 900° C, D = 10^{-6} cm^2/s. If $\beta = 10^{-8}$ cm, then $\Gamma = 10^{10}$/s. That is, each carbon atom changes position about 10 billion times per second.

Near their melting points most fcc and hcp metals have a self-diffusion coefficient of 10^{-8} cm^2/s. Again taking $\beta = 10^{-8}$ cm gives Γ = 10^8/s. Thus in most solid metals near their melting point each atom changes its site 100 million times a second. If this number seems impossibly large, remember that the vibrational frequency (Debye frequency) of such atoms is 10^{13} to 10^{14}/s, so that the atom only changes position on one oscillation in 10^4 or 10^5. Thus, even near the melting point, the great majority of the time the atom is oscillating about its equilibrium position in the crystal.

2.2 MECHANISMS OF DIFFUSION

It is well known from the theory of specific heats that atoms in a crystal oscillate around their equilibrium positions. Occasionally these oscillations become large enough to allow an atom to change sites. It is these jumps from one site to another which give rise to diffusion in solids. The discussion given in the preceding section and most of the kinetic arguments given in this chapter will apply to several or all of the possible diffusion mechanisms. However, to aid the reader in understanding the applicability of the subsequent material, this section will be devoted to cataloguing the mechanisms which are thought to give rise to diffusion in crystalline solids.

Interstitial Mechanism. An atom is said to diffuse by an interstitial mechanism when it passes from one interstitial site to one of its nearest-neighbor interstitial sites without permanently displacing any of the matrix atoms. Figure 2-2 shows the interstitial sites of an fcc lattice. An atom would diffuse by an interstitial mechanism in this lattice by jumping from one site to another on this sublattice of interstitial points. (Note that these sites also form an fcc lattice.)

Consider the atomic movements which must occur before a jump can occur. Figure 2-3 shows an interstitial atom in the (100) plane of

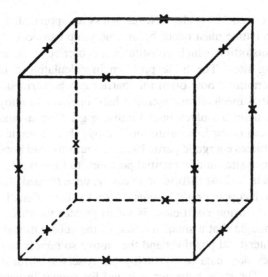

Fig. 2-2—x indicates the interstitial sites in an fcc cell.

a group of spheres packed into an fcc lattice. Before the atom labeled 1 can jump to the nearest-neighbor site 2 the matrix atoms labeled 3 and 4 must move apart enough to let it through. Actually if 1 rises out of the plane of the paper slightly as it starts toward 2, there is a par-

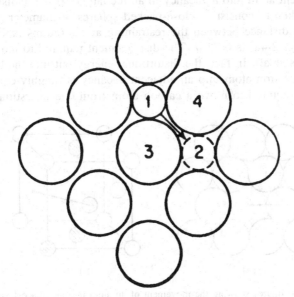

Fig. 2-3—(100) plane in fcc lattice showing path of interstitial solute diffusing by interstitial mechanism.

tially formed channel available. Nevertheless an appreciable local dilatation of the lattice must occur before the jump can occur. It is this dilatation or distortion which constitutes the barrier to an interstitial atom changing sites. The basic problem in calculating a jump frequency is determining how often this barrier can be surmounted.

The interstitial mechanism described here operates in alloys where the solute normally dissolves interstitially, e.g., C in α- and γ-iron. In addition it can occur in substitutional alloys. For example in radiation damage studies energetic particles, e.g. neutrons, can knock atoms off normal lattice sites into interstitial positions to form what are called "self-interstitials." These diffuse quite easily, once formed. As another example, though copper or gold atoms dissolve substitutionally in lead, their average diffusion coefficient is much greater than that for lead atoms. It is thought that a small fraction of the substitutional gold atoms go into interstitial positions and then move so rapidly through the lattice that they dominate in producing the observed diffusion of the solute in lead. The same behavior is found for copper in germanium.

Vacancy Mechanism. In all crystals some of the lattice sites are unoccupied. These unoccupied sites are called vacancies. If one of the atoms on an adjacent site jumps into the vacancy, the atom is said to have diffused by a vacancy mechanism.

Figure 2-4 shows the nature of the constriction which inhibits motion of an adjacent atom into a vacancy in an fcc lattice. If the undistorted lattice is taken to consist of close-packed spheres of diameter d, the equilibrium distance between the restraining atoms (atoms labeled 1 and 2 in Fig. 2-4a) is 0.73 d. The displacement required to move an atom is thus small. In fact, the distortional energy put into the lattice in moving an iron atom into an adjacent vacancy is roughly equal to the energy required to move a carbon atom from one interstitial site

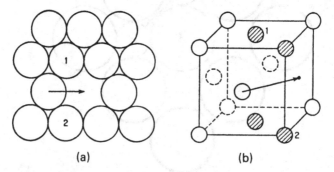

(a)   (b)

Fig. 2-4—Two figures showing the movement of an atom into an adjacent vacancy in an fcc lattice. (a) A close-packed plane of spheres. (b) A unit cell showing the four atoms (shaded) which must move for the indicated jump to occur.

to another in the same fcc phase. The reason that iron diffuses so much more slowly than carbon is that while each carbon atom always has many vacant nearest-neighbor interstitial sites, the fraction of vacant iron sites is very small, and each iron atom must wait an appreciable period before a vacancy becomes available.

In a bcc lattice the barrier for the jump of an atom into a vacant nearest neighbor site is more complex. Fig. 2-5 represents the extended barrier of two sets of triangular barriers. The vacancy mechanism is thought to be the mechanism of self diffusion for all pure metals and for essentially all substitutional solutes in alloys. It also is found in ionic compounds and oxides.

Interstitialcy and Crowdion Mechanisms. Solute atoms which go into solution in metals as interstitials are appreciably smaller than the matrix atoms and, as discussed above, diffuse by the interstitial mechanism. If a relatively large atom such as a solvent atom gets into an interstitial position, how will it move? It will produce a very large distortion if it jumps from one interstitial site to a neighboring interstitial site. Jumps which produce very large distortions occur infrequently, so another diffusion mechanism which produces less distortion could predominate.

One jump process which gives less distortion is the interstitialcy mechanism. Consider the interstitial atom shown in Fig. 2-6. It is said to diffuse by an interstitialcy mechanism if it pushes one of its nearest-

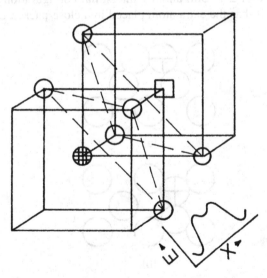

Fig. 2-5 — Saddle point barrier for the darkened atom jumping to the vacancy indicated in a bcc lattice. Note double maxima in energy vs. distance for the jump.

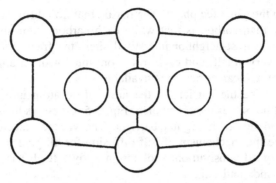

Fig. 2-6—(100) plane of fcc lattice with one atom on an interstitial site.

neighbor atoms into an interstitial position and occupies the lattice site previously occupied by the displaced atom. The distortion involved in this displacement is quite small, so it can occur with relative ease. The mechanism has proved to be the dominant one for the diffusion of silver in AgBr (Chap. 5). In this case the silver ion is smaller than the Br, and an interstitial silver ion does not distort the lattice unduly.

In the case of pure fcc metals the atoms are all the same size, and the distortion associated with the configuration shown in Fig. 2-6 is quite large. It has been shown that for Cu, and probably for all fcc metals, the accommodation of the extra (interstitial) atom in the man-
· ner shown in Fig. 2-7. Still another interstitial configuration is called the crowdion. It has the extra atom placed in a close-packed direction,

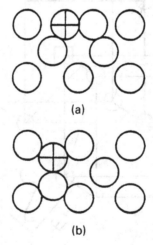

Fig. 2-7—(100) plane of fcc lattice with two atoms sharing one site. The difference between (a) and (b) is an interstitialcy jump.

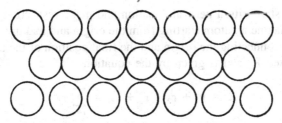

Fig. 2-8 — (111) plane of fcc lattice showing a crowdion. (Note extra atom in middle row.)

thus displacing several atoms from their equilibrium position (see Fig. 2-8). This configuration resembles an edge dislocation in that its distor- tion is spread out along a line, it can glide in only one direction, and the energy to move it is quite small.

With this multiplicity of configurations for an interstitial atom it is well to point out that an interstitial atom means only that there is one more atom than there are sites in a given small region. Similarly, that a vacancy need not mean that a particular site is vacant but that the region contains one fewer atom than sites.

2.3 RANDOM-WALK PROBLEM

After cataloguing the possible diffusion mechanisms, we turn next to the problem of relating these atomic jumps to the observed macroscopic diffusion phenomena. It has already been shown that near the melting point of many metals each atom changes sites roughly 10^8 times per second. Over the period of hours or days, the number of jumps becomes astronomical. These jumps are made in all directions and follow no particular pattern. Our problem is to take this welter of jumps and calculate the mean distance an atom will move from its initial site in n jumps. A first impression might be that the problem is insoluble, owing to the randomness of the atoms's jumps, and indeed the exact distance cannot be calculated for a particular atom. However, precisely because of the random nature of the process and the large number of jumps, it is relatively easy to calculate the average distance that a group of atoms will have migrated from their initial sites. This kind of problem is called a "random-walk" problem, and diffusion in crystalline solids is only one application of a broad group which includes the flipping of coins, the structure of polymers, and the theory of galaxies.[3]

---

[3]For further reading on the random walk problem, G. Gamov, *One, Two, Three . . . Infinity*, Viking Press (1947), Chap. 8, gives a very readable introduction. J. Manning, *Diffusion Kinetics for Atoms in Crystals*, Van Nostrand, 1968, gives a more detailed discussion of its use in diffusion problems.

We will start with a general equation and make restrictions only as needed. Imagine an atom starting from the origin and making n jumps. The vector connecting the origin and the final position of the atom will be designated $\mathbf{R}_n$ and is given by the equation

$$\mathbf{R}_n = \mathbf{r}_1 + \mathbf{r}_2 + \mathbf{r}_3 + \ldots = \sum_{i=1}^{n} \mathbf{r}_i \tag{2-3}$$

where the r_i are vectors representing the various jumps. To obtain the magnitude of R_n, we square both sides of Eq. (2-3).

$$\mathbf{R}_n \cdot \mathbf{R}_n = R_n^2 = \mathbf{r}_1 \cdot \mathbf{r}_1 + \mathbf{r}_1 \cdot \mathbf{r}_2 + \mathbf{r}_1 \cdot \mathbf{r}_3 + \ldots + \mathbf{r}_1 \cdot \mathbf{r}_n$$

$$+ \mathbf{r}_2 \cdot \mathbf{r}_1 + \mathbf{r}_2 \cdot \mathbf{r}_2 + \mathbf{r}_2 \cdot \mathbf{r}_3 + \ldots + \mathbf{r}_2 \cdot \mathbf{r}_n$$

$$\cdots\cdots\cdots\cdots\cdots\cdots\cdots\cdots\cdots\cdots\cdots\cdots\cdots\cdots \tag{2-4}$$

$$+ \mathbf{r}_n \cdot \mathbf{r}_1 + \mathbf{r}_n \cdot \mathbf{r}_2 + \mathbf{r}_n \cdot \mathbf{r}_3 + \ldots + \mathbf{r}_n \cdot \mathbf{r}_n$$

We can rewrite this array as a series of sums in which the first sum is the sum of the diagonal terms, $\mathbf{r}_i \cdot \mathbf{r}_i$. The second sum will consist of all the terms $\mathbf{r}_i \cdot \mathbf{r}_{i+1}$ and $\mathbf{r}_i \cdot \mathbf{r}_{i+1}$. There are $n - 1$ of each of these terms, and they can be said to lie along the semidiagonals of Eq. (2-4). Since $\mathbf{r}_i \cdot \mathbf{r}_{i+1}$ equals $\mathbf{r}_{i+1} \cdot \mathbf{r}_i$, these two sums can be combined. Proceeding in this manner gives

$$R_n^2 = \sum_{i=1}^{n} \mathbf{r}_i \cdot \mathbf{r}_i + 2 \sum_{i=1}^{n-1} \mathbf{r}_i \cdot \mathbf{r}_{i+1} + 2 \sum_{i=1}^{n-2} \mathbf{r}_i \cdot \mathbf{r}_{i+2} + \ldots$$

$$= \sum_{i=1}^{n} r_i^2 + 2 \sum_{j=1}^{n-1} \sum_{i=1}^{n-j} \mathbf{r}_i \cdot \mathbf{r}_{i+j} \tag{2-5}$$

To put this in the form we shall finally work with, note that by definition $\mathbf{r}_i \cdot \mathbf{r}_{i+j} = |r_i||r_{i+j}|\cos \theta_{i,i+j}$ where $\theta_{i,i+j}$ is the angle between the two vectors. Substituting this relation in Eq. (2-5) gives

$$R_n^2 \sum_{i=j}^{n} r_i^2 + 2 \sum_{j=1}^{n-1} \sum_{i=1}^{n-j} |r_i||r_{i+j}| \cos \theta_{i,i+j} \tag{2-6}$$

Note that in the derivation of this equation no assumptions have been made concerning: (1) the randomness of the jumps, (2) the lengths of the successive jumps, (3) the allowable values of $\theta_{i,i+j}$, or (4) the number of dimensions in which the atom is jumping. We shall proceed to make assumptions about these and calculate an average value of R_n^2.

The problem of primary interest is that of diffusion in a crystalline solid. For crystals with cubic symmetry all the jump vectors will be equal in magnitude, and Eq. (2-6) can be written

$$R_n^2 = nr^2 + 2r^2 \sum_{j=1}^{n-1} \sum_{i=1}^{n-j} \cos \theta_{i,i+j}$$

$$= nr^2 \left(1 + (2/n) \sum_{j=1}^{n-1} \sum_{i=1}^{n-j} \cos \theta_{i,i+j} \right) \qquad (2\text{-}7)$$

This equation gives R_n^2 for one particle after n jumps. To obtain the average value of R_n^2, we must consider many atoms, each of which has taken n jumps. The quantity nr^2 will be the same for each flight, but the values of R_n^2 will be different, and the differences will arise from the differences in the magnitudes of the double sums. The average value of R_n^2 can be obtained by adding the various R_n^2 and dividing the sum by the number of atoms involved. The result can be written

$$\overline{R_n^2} = nr^2 \left(1 + (2/n) \overline{\sum_{j=1} \sum_{i=1} \cos \theta_{i,i+j}} \right) \qquad (2\text{-}8)$$

If each jump direction is independent of the direction of the jumps which preceded it and each jump vector and its negative are equally probable, then positive and negative values of any given $\cos \theta_{i,i+j}$ will occur with equal frequency, and the average value of the term involving the double sum will be zero. When this is true

$$\overline{R_n^2} = nr^2 \qquad (2\text{-}9)$$

or

$$\sqrt{\overline{R_n^2}} = \sqrt{n}\, r \qquad (2\text{-}10)$$

Two aspects of this result are particularly noteworthy. The first is the extreme simplicity of the equation. The second is the fact that the mean displacement (actually the root-mean-square displacement) is proportional to the square root of the number of jumps.

As a simple example of the step between Eqs. (2-8) and (2-9), consider the case of a single atom jumping back and forth along a line. The values of $\cos \theta_{i,i+j}$ are now either $+1$ or -1 since the angle between any two jump vectors will be either 0 or 180. If we consider many lines, each with a particle initially at $x = 0$ as in Fig. 2-9a, after n jumps the particles will be distributed at various distances from $x = 0$ as shown in Fig. 2-9b. From this figure it is apparent that R_n^2 for the various particles differs appreciably. However, since values of $\cos \theta = +1$ and $\cos \theta = -1$ are equally probable, $2/n$ times the average value of $\cos \theta$ for all jumps will be much less than 1.

Although the number of jumps and number of atoms are very small, let us consider in more detail the case depicted in Fig. 2-9. To obtain the path of each atom, a coin was flipped 16 times. A head represented

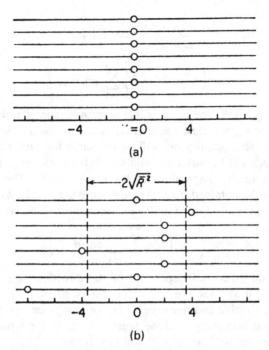

Fig. 2-9 — (a) Initial distribution of atoms, one to a line. (b) Final distribution after each atoms made 16 random jumps. $\sqrt{\bar{R}^2}$ is the calculated root mean square of the points shown.

a jump to the right; a tail, a jump to the left. In this way successive jumps were independent of each other, and jumps to the left and to the right were equally probable. The root-mean-square displacement of this group of atoms is found to be 3.2 units. Since there were 16 jumps per atom, Eq. (2-9) predicts a value of 4.0, which is in reasonable agreement considering the very small number of jumps involved.

To give a better understanding of the effect of the random movement of atoms for very large values of n, consider the case of carbon diffusing in γ-iron. At an average carburizing temperature (950° C) carbon atoms make 10^{10} jumps per second. Taking the jump distance to be 0.1 nm, in one second each carbon atoms travels a total distance of 1 m and a net distance of 10^{-3} cm. After the 10^4 s (about 3 hr) of an average carburizing treatment, the mean penetration is 0.10 cm, while the total distance traveled by the atom is 10 km. It is thus obvious that the net displacement of each atom is extremely small compared to the total distance it travels.

Random Walk in FCC Lattice. As a second example of what is involved in going from Eq. (2-8) to Eq. (2-9), consider the case of a

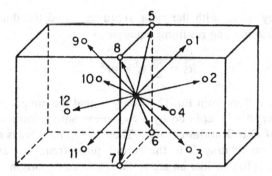

Fig. 2-10—The arrows show 12 possible jump vectors in an fcc lattice.

vacancy diffusing in an fcc lattice.[4] Figure 2-10 shows the orientation of the 12 possible jump vectors for this lattice. All of these vectors are equivalent so that each will occur with the same frequency if an average is taken over many vacancies and a very large number of jumps. For this reason the mean value of the quantity $r_i \cdot r_{i+j}$, i.e. $r_i \cdot r_{i+j}$ for $i \neq j$, is independent of the vector chosen for r_{i+j}. Furthermore, the very large number of vectors to be dotted into r_i, will consist of the 12 vectors of Fig. 2-10 in equal proportions. Thus the mean value of the double sum in Eq. (2-5) will be zero if

$$\sum_{j=1}^{12} r_i \cdot r_j = r^2 \sum_{j=1}^{12} \cos \theta_{ij} = 0 \qquad (2\text{-}11)$$

where the summation of j ranges over the 12 jump vectors of Fig. 2-10. It can be seen in Fig. 2-10 that for any particular jump vector there is another jump vector equal to the negative of that vector. For example, $r_5 = -r_7$. Therefore

$$r_i \cdot r_7 + r_i \cdot r_5 = r_i \cdot r_7 - r_i \cdot r_7 = 0$$

Pairing up each of the 12 vectors with its negative in this way, we see that Eq. (2-11) is satisfied, and so again we obtain

$$\overline{R_n^2} = nr^2 \qquad (2\text{-}9)$$

A similar argument can be given for the case of liquids or gases. The two main differences between this case and crystals is that all values of $\theta_{i,i+j}$ are possible instead of a discrete set and all values of r are possible (although in any given system the values will cluster about some mean). If the jumps, or flights, r_j and $-r_j$ are equally

---

[4]The proof given here applies equally well for any other cubic lattice in which all atoms are on equivalent sites.

probable, they occur with the same frequency, and the double sum again goes to zero. The resulting equation is

$$\overline{R_n^2} = \sum_{i=1}^{n} r_i^2 = nr^2 \qquad (2\text{-}12)$$

This equation differs from Eq. (2-9) only in that the unique jump distance of the crystal is replaced by a root-mean-square jump distance.

Relation of D to Random Walk. There are several ways of deriving the equation relating D to the atomic jump frequency and jump distance. One is to consider an atom starting from the origin, as in the preceding section, but instead of calculating R_n^2, to calculate the probability that the atom has moved from the origin to a distance between r and $r + dr$ after n jumps. This probability can also be calculated by solving the same problem using Fick's laws. In addition to giving the desired equation for r, this procedure shows that the two approaches give identical answers. The main drawback to this approach for our purposes is that the probability problem is complicated, though not difficult.[5]

A second, simpler approach is to consider the same problem but to solve only for the mean value of R_n^2 using the two techniques. It is shown in one of the problems at the end of this chapter that, if $C = n/t$ and α is the jump distance[6]

$$\overline{R_n^2} = n\alpha^2 = 6Dt \qquad (2\text{-}13)$$

so that

$$D = (1/6)\Gamma\alpha^2 \qquad (2\text{-}14)$$

This equation differs from Eq. (2-2) in that the jump distance in three dimensions, α, may not equal the distance between planes, β.

A simpler derivation of Eq. (2-14) will now be given working with a particular mechanism in a particular three-dimensional lattice. Since most of the self-diffusion studies have been made on fcc metals, the diffusion of a tracer in a pure fcc metal by a vacancy mechanism is considered.

If Γ is the average number of jumps per second for each tracer atom and n_1 is the number of tracer atoms on plane 1, $n_1\Gamma\delta t$ of the tracer atoms on plane 1 will jump in the short period δt. The quantity $\Gamma\delta t$ will be proportional to the number of nearest-neighbor sites, to the

---

[5] A solution of this type is given in the first few pages of B. S. Chandrasekhar, *Revs. Modern Phys.*, *15* (1943) 1. This article is also reprinted in N. Wax (ed.), *Selected Papers on Noise and Stochastic Processes*, Dover Publications, New York, 1954.

[6] Or see M. N. Barber, B. W. Ninham, *Random and Restricted Walks*, Gordon & Breach, (1970) Chap. 8.

probability that any given neighboring site is vacant (p_v), and to the probability that the tracer will jump into a particular vacant site in δt, namely $w\delta t$.[7] Thus we can write

$$\Gamma\,\delta t = 12 p_v\, w\,\delta t \tag{2-15}$$

Since only 4 of the 12 nearest neighbors are on the plane the given atom is jumping to (see Fig. 2-10), the flux per unit area from plane 1 to plane 2 is

$$J_{12} = 4n_1 p_{v2} w_{12}$$

where subscripts have been added to emphasize the planes involved; for example, p_{v2} is the probability that any site on plane 2 is vacant. Similarly the reverse flux is given by

$$J_{21} = 4n_2 p_{v1} w_{21}$$

In alloys, w and p_v will change with composition and thus give rise to a variation of D with composition. However, all isotopes of a metal are assumed to act the same, so that in a pure metal $w_{12} = w_{21}$ and $p_{v1} = p_{v2}$. Combining these last two equations and noting that the distance between planes $\beta = a_o/2$, gives for the net flux

$$J = 4p_v\, w(n_1 - n_2) = 4p_v\, w(a_o/2)(c_1 - c_2) \tag{2-16}$$

or substituting

$$c_1 - c_2 = -\frac{a_o}{2}\frac{\partial c}{\partial x}$$

we get

$$J = -a_o^2 p_v\, w(\partial c/\partial x)$$

From the assumed equivalence of all isotopes of the same metal, it follows that p_v will be equal to the fraction of sites vacant, or N_v. The desired equation is then

$$D = a_o^2 N_v\, w \tag{2-17}$$

The calculation of D in a pure fcc metal is thus reduced to the problem of calculating the mole fraction of vacancies and the jump frequency of an atom into an adjacent vacancy. Conversely, if we measure D, since a_o is known, we can calculate $N_v w$; or knowing N_v, we can calculate w.

In a derivation similar to that which led to Eq. (2-17), it is possible to derive an equation for an interstitial solute in a binary alloy. If the

---

[7] If Γ_v is the number of jumps a vacancy make per second, then $12w = \Gamma_v = \Gamma/p_v$.

solution is very dilute, w is independent of composition, and the mole fraction of vacant interstitial sites is essentially unity. Thus for a dilute alloy, D for the interstitial element is

$$D = \gamma a_o^2 w \qquad (2\text{-}18)$$

where γ is a geometric constant derivable from Eq. (2-14).

In closing, note that both Eqs. (2-17) and (2-18) can also be derived from Eq. (2-14) by substituting $\alpha = a_o/\sqrt{2}$ and $\Gamma = 12wN_v$. The advantage of the derivation given above is its relative simplicity and the ease with which it can be extended to the cases in which p_v and w vary with composition, or the drift that occurs when $w_{12} \neq w_{21}$. However, once the assumptions leading to Eqs. (2-17) and (2-18) are clearly in mind, it is usually easier to work with Eq. (2-14).

2.4   CALCULATION OF D

Our study of the atomistic processes contributing to diffusion has led to Eq. (2-17). From this point on, our understanding of D will increase in proportion to our understanding of w and N_v. Thus the calculation of these terms is one of the basic problems in the atomic approach to diffusion. In this section we shall review the methods of evaluation, trying to emphasize the assumptions and approximations involved. The discussion to be given is strictly applicable only to self-diffusion in pure metals or interstitial diffusion in very dilute, binary alloys. The changes needed in an extension to substitutional alloys is discussed in Chap. 4.

Equilibrium Concentration of Vacancies. To gain a better understanding of D for a vacancy mechanism, we consider first the problem of how many vacancies will be present in a pure metal and how this concentration will change with temperature. The most important concept to be grasped here is the increase in entropy which results from the mixing of two pure components. A plausibility argument for this increase in entropy can be seen from the following. If a drop of ink is placed in a glass of water, the mixture will ultimately become uniformly tinted. An explanation for this homogenization might be found in Fick's laws, but it could also be found in a basic thermodynamic requirement for equilibrium; namely, that for equilibrium the entropy of any isolated system will be maximum. Thus, this homogenization or mixing of the ink-water mixture must correspond to an increase of the entropy of the mixture.

To be more quantitative, if an ideal solution is formed upon the addition of component 1 to component 2, the equation for the increase of entropy per mole of solution is

$$S_{mix} = -R[(1 - N_1)ln(1 - N_1) + N_1 ln(N_1)] \qquad (2-19)$$

where N_1 is the mole fraction of component 1. This equation is plotted in Fig. 2-11. It is seen that the entropy per mole of any mixture is greater than that of the pure components. To apply Eq. (2-19) to the study of vacancies in metals, consider a large piece of pure metal with no sites vacant. If several vacancies are taken from the surface and mixed throughout the metal, the increase in the entropy, per mole of solution δS_{mix}, is

$$\delta S_{mix} = \frac{dS_{mix}}{dN_v} \delta N_v = -R \, ln\left(\frac{N_v}{1 - N_v}\right) \delta N_v \qquad (2-20)$$

where δN_v is the change in the mole fraction of vacancies. In the limit of $N_v \to 0$ it can be seen that $\delta S_{mix}/dN_v \to \infty$. That is, the increase in entropy per vacancy added is extremely large for the first few vacancies, but it continually decreases from its initial, infinite value. It follows that at equilibrium there will always be some vacancies in a piece of metal. To calculate just what the equilibrium value of N_v will be, we use the fact that in any isothermal, isobaric system at equilibrium the change in the Gibbs free energy G will be zero for any small displacement. If dn_v additional vacancies are mixed into a mole of a crystal already containing the concentration N_v of vacancies, the change in G will be

$$\delta G = H_v \frac{\delta n_v}{N} - T \frac{\partial S}{\partial N_v} \frac{\delta n_v}{N} \qquad (2-21)$$

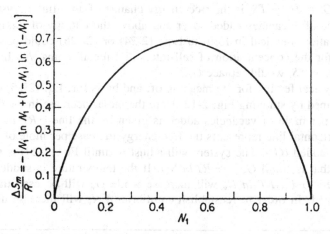

Fig. 2-11—S_{mix} is the entropy increase upon forming one mole of an ideal solution from the pure components. The slope of the curve is infinite at the limits of $N_1 = 0$ and $N_1 = 1$.

where N is Avogadro's number. H_v/N is the increase in the enthalpy of the crystal per vacancy added and stems from the local changes in the atomic and electronic configurations of the crystal when a vacancy is introduced. The increase in the entropy of the lattice per vacancy added, $(\partial S/\partial N_v)(1/N)$, arises from the ideal entropy of mixing as given by Eq. (2-20) and a second part which stems primarily from the change in the vibrations of the atoms when a vacancy is introduced. This second contribution is designated S_v/N per vacancy.[8] Substituting these terms in Eq. (2-21) gives the equation

$$\delta G = [H_v - TS_v + RT \ln N_v/(1 - N_v)](\delta n_v/N) \qquad (2\text{-}22)$$

Both H_v and S_v will be independent of N_v in very dilute solutions where the vacancies do not interact with one another. (Experiments indicate that in pure metals $N_v < 10^{-4}$, so the solution is indeed very dilute.) Since $N_v \ll 1$, Eq. (2-22) can be written

$$\delta G = [H_v - TS_v + RT \ln N_v](\delta n_v/N) \qquad (2\text{-}23)$$

But, at equilibrium $\delta G = 0$ for any small δn_v. Thus at equilibrium, N_v must have the value given by the equation

$$N_v^e = \exp(S_v/R) \exp(-H_v/RT) \qquad (2\text{-}24)$$

where the superscript e is added to N_v to emphasize that N_v^e is a particular value of N_v instead of a variable. This equation can also be written

$$N_v^e = \exp(-G_v/RT) \qquad (2\text{-}25)$$

where $G_v = H_v - TS_v$ is the free-energy change of an infinite crystal, per mole of vacancies added, over and above the entropy of mixing. An equation identical in form to Eqs. (2-24) or (2- 25) could be obtained for the concentration of self-interstitial metal atoms, N_i. In it, $G_i = H_i - TS_i$ would replace G_v.

A physical feeling for the meaning of, and basis for, Eq. (2-25) can be obtained by studying Fig. 2-12. Here the molar decrease in the free energy per mole of vacancies added is given by the line $-RT \ln N_v$. The horizontal line represents the free-energy increase per mole of vacancies added (G_v). The system will adjust N_v until Eq. (2-24) is satisfied, that is, until $G_v = -RT \ln(N_v)$. If the temperature is suddenly increased to T_2, $RT \ln N_v$ will increase while G_v will be essentially unchanged. In order to reestablish equilibrium, N_v will increase until

---

[8]Strictly speaking, S_v is a parameter which when added to the ideal entropy of mixing gives the observed entropy effect. In solution chemistry S_v would be called an excess entropy of mixing.

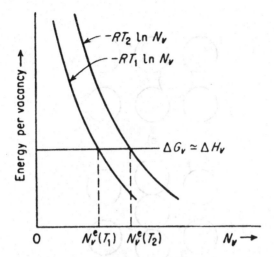

Fig. 2-12—$N_v^e(T)$, the equilibrium concentration of vacancies at temperature T, is attained when $T(dS_{mix}/dN_v) = -G_v$. The variation of both of these quantities with N_v is shown, at two temperatures.

$-RT\,ln(N_v)$ again equals G_v. For vacancies in gold, $H_v = 191$ kJ/mol ($=1.0$ eV per vacancy). Thus in the temperature range of 900–1000° C, N_v in gold will roughly double with a 90° C increase in temperature. Often authors write $N_v = \exp(-H_v/RT)$, omitting the term including S_v. This is not correct, but few data available indicate that $\exp(S_v/R) < 10$. The omission of this term often gives an adequate approximation and avoids the problems of discussing S_v. Taking $S_v = 0$ and $H_v = 23.0$ kcal/mol, we get $N_v = 10^{-4}$ at 980° C.

Calculation of the Jump Frequency w. The second unknown quantity which enters into D is w, the frequency with which an atom will jump into an adjacent, vacant site. The calculation of w, or even its temperature dependence, from our fundamental knowledge of the forces between atoms and reaction kinetics is very difficult. Actually, our present knowledge is such that any calculation from fundamentals cannot give a real check on experimental results. The main purpose in such a theoretical study is to develop greater insight into the factors which determine w and thereby D.

The atom movements required for an atom to jump are shown schematically in Fig. 2-13; (a) and (c) show the initial and final states, while (b) shows the midway configuration referred to as the activated state. There are two separate requirements to be met before the group of atoms can go from (a) to (c). First, the diffusing atom must be moving to the right far enough to carry it into the adjacent site; and second, the two restraining atoms must simultaneously move apart a

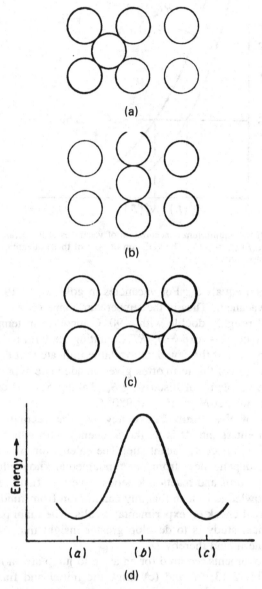

Fig. 2-13—(a), (b), and (c) are schematic drawings showing the sequence of configurations when an atom jumps from one normal site to a neighboring one. (d) shows how the free energy of the entire lattice would vary as the diffusing atom is reversibly moved from configuration (a) to (b) to (c).

great enough distance to let the diffusing atom through. Whenever these two steps occur at the same time, the diffusing atom will change sites.

The most common method of calculating w ignores the detailed atomic movements involved and uses statistical mechanics to calculate the concentration of "activated complexes," or regions containing an atom midway between two equilibrium sites. The number of atoms diffusing per second is then obtained by multiplying the number of activated complexes (n_m) by the average velocity of the atoms moving through this midpoint $\bar{v}$, divided by the width of the barrier or midpoint δ. From this number jumping per second, it is shown that the average jump frequency per atom is $w = N_m\bar{v}/\delta$, where N_m is the mole fraction of activated complexes. The treatment of this problem in a rigorous manner is very difficult since it is a many body problem, and the vibrations leading to a site change are no longer harmonic.[9] The simplified treatment given here is chosen to make the basic assumptions apparent.

The diffusing atom shown schematically in Fig. 2-13b is said to be at the saddle point. Throughout the crystal there will always be atoms entering this configuration as well as leaving it. To calculate the number of atoms at the saddle point at any instant, it is necessary to know the increase in the Gibbs free energy of a region when an atom in it moves from a normal site to the saddle-point position, G_m. Zener[10] suggested that this free-energy change could be visualized in the following thought experiment. If the diffusion direction is defined as the x axis, we constrain the atom so that it can execute its normal vibration only in the yz plane. The atom is then slowly (reversibly) moved from its initial site to the saddle point, allowing the surrounding atoms to continuously readjust their positions. The work done in this reversible, isothermal process, at constant pressure, is just equal to the change in Gibbs free energy for the region (G_m). This can be written

$$G_m = H_m - TS_m \tag{2-26}$$

It is assumed that G_m has all the properties possessed by G_v of Eq. (2-25). Given G_m, the equilibrium mole fraction of atoms in the region of the saddle point N_m can be calculated using a treatment essentially the same as was used in obtaining the equation for N_v^e, that is, Eq (2-25). Instead of mixing into the lattice vacancies which increase the free energy by G_v per mole of vacancies, we mix in activated complexes which increase the free energy by G_m per mole of complexes.

[9]For a more detailed discussion of the problem see Chap. 7 of C. P. Flynn, *Point Defects & Diffusion*, Clarendon-Oxford Press, (1972).

[10]C. Zener, in W. Schockley (ed.), *Imperfections in Nearly Perfect Crystals*, p. 289, John Wiley & Sons, Inc., New York, 1952, or C. P. Flynn, *Point Defects & Diffusion*, Clarendon-Oxford Press, (1972), pp. 335–7.

The ideal entropy of mixing is the same for vacancies and complexes so, at equilibrium, n_m out of N atoms will be in the neighborhood of a saddle point at any instant and

$$n_m/N = N_m = \exp[(-H_m + TS_m)/RT] = \exp(-G_m/RT) \quad (2\text{-}27)$$

In the equation $w = N_m \bar{v}/\delta$, simple dimensional analysis shows that $\bar{v}/\delta$ is a frequency. This is the frequency v with which the atoms at the saddle point go to the new site. A more complete treatment shows that v is of the order of the mean vibrational frequency of an atom about its equilibrium site. Thus, of N atoms $n_m v$ will jump from one site to a given vacant neighbor site per second. If this is true, the average jump frequency for any given atom will be

$$vn_m/N = w = v \exp(-G_m/RT) \quad (2\text{-}28)$$

A particularly simple interpretation of Eq. (2-28) is to think of it as the frequency with which an atom vibrates in a given diffusion direction v times the probability that any given oscillation will move the atom to an adjacent site in that direction, $\exp(-G_m/RT)$. The precise definition of v is one of the more difficult aspects of a rigorous theory. However, it is usually taken equal to the Debye frequency.

Equations for D. Empirically it is found that D can be described by the equation

$$D = D_o \exp(-Q/RT) \quad (2\text{-}29)$$

where D_o and Q will vary with composition but are independent of temperature. Experimentally D_o and the activation energy Q are obtained by plotting $\ln D$ versus $1/T$. The slope of this plot gives

$$\frac{d \ln D}{d\, 1/T} = -\frac{Q}{R}$$

while $\ln D_o$ is given by the intercept at $1/T = 0$.

An alternate equation for D in the case of interstitial diffusion can be obtained by substituting Eq. (2-28) for w in Eq. (2-18). This gives

$$D = [\gamma a_o^2 v \exp(S_m/R)] \exp[-H_m/RT] \quad (2\text{-}30)$$

Comparing this with Eq. (2-29) shows that the first term in parentheses is equal to D_o and that Q equals the quantity H_m.

For diffusion by a vacancy mechanism in a pure metal, Eqs. (2-28), (2-24), and (2-17) give

$$D = \left[a_o^2 v \exp\left(\frac{S_v + S_m}{R} \right) \right] \exp \left(\frac{-H_v - H_m}{RT} \right) \quad (2\text{-}31)$$

The term in square brackets is again D_o, while Q is the sum of H_m and H_v. Since Q is seen to be made up of enthalpy terms in both cases.

The entropy terms $S_v + S_m$ from either Eq. (2-30) or Eq. (2-31) can be evaluated from the known value of D_o, a_o^2, γ, and an assumed value of ν. As was pointed out above, ν is usually taken to be the Debye frequency for a pure metal. In the case of interstitial atoms, ν can be estimated by assuming that the potential-energy curve of the atom varies sinusoidally along the diffusion path and its maximum value is H_m.[11] In either case, the value of the entropy term obtained depends on the value of ν assumed. In view of the vagueness as to what ν is to be used, the S_m cannot be determined with precision. However, as will be seen in the next section, the evaluation of even an approximate value can be quite helpful in checking experimental results.

2.5 CALCULATION OF H AND S FROM FIRST PRINCIPLES

There has been a fruitful interaction between the theoretical calculations and experiment in this area. Historically one of the first questions was whether diffusion in fcc noble metals like copper occurs by the exchange of two adjacent atoms or by a vacancy mechanism. Later with the advent of big computers and studies of radiation damage, the theoretical models provided insights into the formation and motion energies of self-interstitials, as well as small defect clusters. Most recently the models have been used to study possible mechanisms of diffusion in grain boundaries.

The models used on the large computers to calculate defect energies starts with an assumed energy function. In this function the energy of the lattice is described as a function of the relative position of all of the atoms in the lattice. This is done with terms arising from the two-body forces between the atoms plus a contribution from changes in electronic structure and volume changes of the crystal.[12]

We shall outline the calculations which have been made using the models of solid-state physics. The actual calculations are beyond the scope of this book. Nevertheless, by reviewing the models and the results, the student will obtain a feeling for the physical effects which contribute to H and S.

Calculation of H_m. The short range interactions between atoms in noble metals and transition metals is determined largely by the repulsion of the filled electron shells, or ion cores, of the atoms. To cal-

---

[11]C. Wert, C. Zener, *Phys. Rev.*, 76, (1949) 1169.

[12]R. A. Johnson, in *Diffusion*, ASM, Metals Park, OH, 1973, pp. 25–46.

culate the energy of activation for an atom jumping into a vacancy, a geometry similar to Fig. 2-13b is used for the activated state. Thus an atom is placed at the saddle point, and the surrounding ions and electrons are allowed to relax to this new configuration. Consideration of the geometry shows that the vacancy has been divided into two equal halves. Thus to a first approximation there is no change in the energy of the electrons, and the electronic contribution to H_m is zero. However, the diffusing atom has moved appreciably closer to its neighbors in the saddle-point configuration, and the ion-core interaction energy is appreciable. The calculation of this interaction for the atoms which are nearest and next nearest neighbors of the activated atom and vacancy is obtained by allowing the atoms to relax until the sum of all of the energy terms is a minimum.

Calculation of H_v. While the migration energy stemmed primarily from ion core repulsion, the energy to form a vacancy stems primarily from the change in energy of the free electrons in the metal. The discussion given here will deal with the metals copper, silver and gold, though it should be similar for transition metals.

To establish a model for calculating H_v, we take advantage of the fact that the enthalpy of the crystal depends only on the number of vacancies present and not on the mechanism by which they were produced. For this reason the conceptual procedure used here to form a vacancy need bear no resemblance to how the vacancies are actually formed in the real crystal. We consider the metal, e.g. copper, to consist of ions with a charge of $+1$, arranged in a gas of electrons. If a neutral atom is removed from the center of the crystal and placed on a rough area of the surface, there is no change in surface area, but there is an increase in the volume. This volume increase decreases the average energy of all of the electrons and gives an energy change of -2.8 eV per vacancy.[13,14]

The removal of an atom from the center of the specimen to the surface leaves one atomic volume devoid of charge. The free electrons in the region around this vacant volume will tend to flow into the vacancy, but since there is no positive charge in the vacant site this will increase the electrostatic energy. This can be seen with the aid of Fig. 2-14 where it is assumed that the positive charge density drops sharply at the edge of the vacant site, while the time average of the electron density tails off into the vacant site. The greater the electron penetration of the vacancy, the greater the electrostatic energy of the separated

---

[13]The energy unit eV (electron volt) is convenient for expressing the energy changes in atomic processes. 1 eV per atom = 96.46 kJ/mol.
[14]The energy changes given here are those of F. Fumi, *Phil. Mag.*, *46* (1955) 1007.

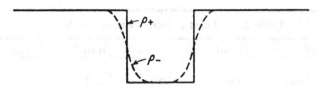

Fig. 2-14—Model for positive and negative charge density distribution (ρ_+ and ρ_-) around a vacancy.

positive and negative charge. However, if this electrostatic energy is minimized by forming a very sharp change in the electron density, shorter wavelengths are required for the electrons and thus higher energies. At equilibrium, the increase in energy is primarily due to the shorter wavelength (higher kinetic energy) of these electrons and is +4.0 eV per vacancy.

Although most of the energy change accompanying the formation of a vacancy is electronic, there is a small contribution from the change in the positions of the ions surrounding the vacancy. These ion cores can be thought of as close-packed spheres which are slightly compressed. If an atom is removed, the surrounding ions will relax into the vacancy, thereby decreasing their energy. This relaxation is small in a close-packed lattice, and the energy decrease from this source is only −0.3 eV per vacancy.

If these three contributions are added together, one obtains a value of 0.9 eV per vacancy = E_v = H_v for copper. The experimentally obtained value of H_v is 1.29 eV (Table 2-1). The values of the energies depend critically on the volume changes assumed, however, the results obtained are of the correct magnitude, and the dominant contribution of electronic terms is clearly indicated.

Calculation of S_v & S_m.[15] It is shown in most texts on statistical mechanics that the Helmholtz free energy of a crystal relative to that at absolute zero can be represented by the equation

$$F = -kT \sum_i ln[1 - \exp(-h\nu_i/kT)]^{-1} \qquad (2\text{-}32)$$

where i is summed over the frequencies of the crystal. The entropy change for some process can then be obtained from Eq. (2-32) by using the thermodynamic equation

$$S = -(\partial F/\partial T)_v \qquad (2\text{-}33)$$

For temperatures well above the Debye temperature, $h\nu_i \ll kT$, and

---

[15]The discussion given here follows that of H. Huntington, G. Shirn, E. Wajda, *Phys. Rev.*, *99* (1955) 1085.

Table 2-1.   Point Defect Energies (in eV)[a]

Metal	$H_v(pa)$	H_m [b]	$H_v(qu)$	$H_v(te)$
bcc				
Mo	3.0	1.50	3.2	
Nb	2.65	1.00		
Ta	2.8	1.45		
V	2.1	1.30		
W	4.0	1.5	3.6	
Fe(par)	1.6	0.92		
fcc				
Al	0.68	0.68	0.66	0.76
Ag	1.12	0.60	1.06	
Au	0.89	0.84	0.94	0.94
Cu	1.29	0.78	1.27	1.17
Fe	1.4	1.26		
Ni	1.78	1.32	1.63	
Pb	0.57	0.59	0.55	
Pd	1.85	0.91		
Pt	1.32	1.37	1.31	

From H. E. Schaefer, *Positron Annihilation*, P. G. Coleman, S. C. Sharma (eds), North Holland (1982) p. 369, except thermal expansion from R. W. Siegel, *J. Nucl. Matl.*, *69&70* (1978) p. 117.
[a]$H_v(pa)$ from positron annihilation, $H_v(qu)$ by quenching, and $H_v(te)$ from thermal expansion.
[b]$H_m = Q_1 - H_v(pa)$.

these two equations give

$$S = -k \sum_i ln(h\nu_i/kT) \qquad (2\text{-}34)$$

If the frequencies of the perfect crystal are designated ν_{io}, and the frequencies after the introduction of a defect as ν_{if}, the entropy change when a defect is introduced is

$$S = k \sum_i ln(\nu_{io}/\nu_{if}) \qquad (2\text{-}35)$$

The summation in Eq. (2-35) extends over all of the vibrational modes of the crystal. Actually solving for all of these modes is too complicated a problem, so Huntington et al. simplified the equations by dividing the lattice into three regions. The first region contains only the nearest neighbors of the defect, and in this region Eq. (2-35) is used. If the defect involved is a vacancy, the force required to slightly displace an atom into the vacancy will be less than that required for the same displacement in a perfect crystal. This means that $\nu_f < \nu_o$, so that the contribution of the atoms in this region would tend to make $S > 0$. If the defect involved is an interstitial atom, the atoms next to the defect will be pushed much closer together, so that $\nu_f > \nu_o$, and this will tend to make $S < 0$.

The second region contains the elastic stress field set up by the defect. Here elasticity theory is applied. As is discussed below under Zener's theory of D_o, the elastic moduli decrease with increasing temperature, so this elastically strained region always makes a positive contribution to S.

The third region contains the rest of the lattice and is only affected by the expansion or contraction required to give zero pressure at the surface. The contribution of this region always reduces the magnitude of the contribution of the first region but does not change its sign.

For the metal copper the values of S calculated for various defects are

$$S_v = 1.5R \quad S_v + S_m = 0.9R \quad S(\text{interstit.}) = 0.8R \quad (2\text{-}36)$$

It is seen that in each case the effect of adding one mole of a defect is to increase the entropy of the crystal by an amount roughly equal to the gas constant R. As a comparison between these calculated values and experimental results, the value of $S_v + S_m$ for copper can be obtained from data on the diffusion coefficient of copper in copper. The experimental values of $D_o = 0.16$ cm^2/s,[16] $\nu = 7 \times 10^{12}$/s, and $a_o = 3.61A$ give $S_v + S_m \simeq 2.8R$. This is considered to be good agreement since the calculated value is only approximate and the value of ν to be taken is also uncertain enough to make up the discrepancy.

2-6.   EXPERIMENTAL DETERMINATION OF H_v, H_m, AND S_v

It is possible to measure H_v, H_m, and S_v experimentally. This work started with efforts to understand the damage introduced into metals by fast neutrons in nuclear power reactors. However that damage is complex and there has been an evolution from the study of these complex annealing processes in non-equilibrium systems to experiments in which the concentration of vacancies can be measured in samples at equilibrium.

Three types of experiments are discussed. The first uses thermal expansion data to obtain values of S_v and H_v. In the second H_v is obtained from the excess vacancies retained in a sample quenched from high temperatures. The third uses positron annihilation to measure H_v.

---

[16]A. Kuper, H. Letaw, L. Slifkin, C. Tomizuka, *Phys. Rev.*, 98 (1955) 1870.

Thermal Expansion. The type of defect which is most important in diffusion studies in metals is thought to be the vacancy. Of the procedures used to determine H_v the simplest to interpret and the only one giving data on S_v comes from the study of thermal expansion.

When a piece of metal is heated, its length L increases. This expansion stems partly from an increase in the distance between lattice planes, but also from the creation of additional vacant sites inside the crystal. The lattice parameter a_o, as determined by X-rays measures only the increase in the average distance between lattice planes. Thus the increase in the atom fraction of sites $\Delta n/n$ will be proportional to the difference between the increase in length of a sample $\Delta L/L$, and the increase in lattice parameter $\Delta a/a$. In general a change in volume dV/V will be three times the change in linear dimension of the same solid, dl/l. Since $\Delta n/n$ is proportional to the change in volume, one arrives at the following expression

$$\Delta n/n = 3(\Delta L/L - \Delta a/a) \qquad (2\text{-}37)$$

If the dominant defect was interstitials and not vacancies then $\Delta n/n$ would be negative. However Eq. (2-37) gives the change in the fraction of atom sites independent of (1) the type of defects (vacancies or interstitials), (2) the degree of lattice relaxation around the sites, or (3) any pairing or clustering of the defects. If both vacancies and interstitials were formed, $\Delta n/n$ would be proportional to the difference between the concentrations of the two defects. In metals, $\Delta n/n$ is positive, as we would expect, and it is assumed that $\Delta n/n$ is due entirely to vacancies.

Equation (2-37) is quite simple, but the experimental measurements required to use it are not. Near the melting point $\Delta n/n = 10^{-4}$. Thus, to measure $\Delta n/n$ with an accuracy of only 1% requires that $\Delta a/a$ and $\Delta L/L$ be measured to within one part in 10^6. This is a nontrivial task at room temperature, and at 700 to 1000° C it becomes a major undertaking. To minimize the effect of errors in temperature measurement, it is necessary to measure $\Delta a/a$ and $\Delta L/L$ on the same specimen at the same time. Careful studies of this type have been reported by Simmons and Balluffi.[17] Their results for aluminum are shown in Fig. 2-15. The difference between the two curves gives the following equation:

$$\Delta n/n = \exp(2.4) \exp(-0.76/kT) \qquad (2\text{-}38)$$

At the melting point of aluminum this gives $\Delta n/n = 9.4 \times 10^{-4}$. The fact that $\Delta n/n$ is positive confirms the belief that vacancies are the

---

[17]R. Simmons, R. Balluffi, *Phys. Rev.*, *119* (1960) 600.

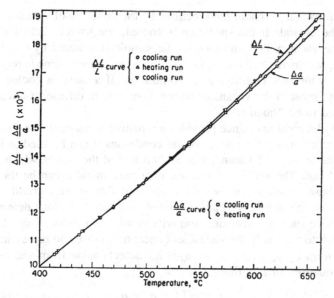

Fig. 2-15—Length change and lattice parameter change vs. temperature for aluminum, taking $\Delta L/L$ and $\Delta a/a$ to be zero at 20° C. The difference between the two lines is directly proportional to the concentration of vacant atomic sites. [From R. Simmons, R. Balluffi, *Phys. Rev.*, *117* (1960) 52.]

dominant type of defect. If there is no interaction between vacancies, they are randomly distributed. In this case $\Delta n/n$ would equal N_v, and Eq. (2-38) would indicate that

$$H_v(te) = 0.75 \text{ eV/vacancy} = 72 \text{ kJ/mole} \quad \text{and} \quad S_v/R = 2.4$$

It is generally agreed that there is a small interaction between the vacancies so that the vacancies will not be randomly distributed. If this is true, the numbers appearing in Eq. (2-38) are not identically equal to S_v/R and H_v,[18] but it is felt that the changes in H_v and S_v/R will be <3%. Other metals have been studied and in each case S_v/R was positive and on the order of unity. For the determination of H_v one can get better data from a wider range of metals using positron annihilation.

Quenching Experiments. If a metal is heated, the new, higher equilibrium concentration of vacancies is established first at dislocations and boundaries, which act as sources. The new concentration

---

[18]Throughout this section G's and H's will be quoted either as the energy change per mole of defects (kJ/mole) or as the energy per defect. This mirrors what the reader will find in the literature and should lead to no confusion since the terminology and units are completely interchangeable. The corresponding molar and atomic entropies are expressed in the dimensionless quantities S/R and S/k, respectively.

then spreads throughout the specimen by the diffusion of vacancies out into the crystal. If the specimen is cooled, the sources act as sinks, and the vacancy concentration of the sample is lowered by the diffusion of vacancies to these sinks. In either case, a finite time is required to reach the new equilibrium concentration. If a metal is cooled very rapidly, most of the vacancies do not have time to diffuse to sinks and are said to be "quenched in."

The electrical resistance provides a sensitive measure of the vacancy concentration, so that under special conditions it can be used to measure the number of vacancies quenched in and the rate at which they anneal out. The specific resistance of a pure metal ρ can be thought of as being made up of two parts, one part due to the thermal oscillations of the lattice $\rho(T)$ and a second part due to various defects in the lattice such as vacancies, impurity atoms, and dislocations. Since we wish to vary only the vacancies concentration we can represent this latter term by $\rho_d + \rho_v$, where ρ_d is all defects aside from vacancies. The equation for ρ is thus

$$\rho = \rho(T) + \rho_d + \rho_v \qquad (2\text{-}39)$$

To measure the resistance change due to some quenching or annealing operation, it is necessary to measure ρ at the same low temperature before and after the operation. This makes $\rho(T)$ the same in all measurements and also makes it small. To make ρ reflect only changes in ρ_v, it is also necessary to keep ρ_d from changing during the cycle. This will be accomplished if the specimen is not contaminated on heating or quenching and if no additional dislocations are introduced by quenching stresses. At the low vacancy concentrations involved, ρ_v will be proportional to N_v, so that if $\rho_d + \rho(T)$ is the same before and after a cycle, we shall have

$$\Delta\rho = \Delta\rho_v = \alpha \, \Delta N_v \qquad (2\text{-}40)$$

Using a resistance bridge, the electrical resistance of a metal specimen can be measured with great precision. Experimentally this means that if a pure metal is held at the temperature of liquid nitrogen [so $\rho(T)$ is small], a change in vacancy concentration will show up as an easily and accurately measurable change in ρ.

Both H_v and H_m can be obtained from quenching experiments. As the specimen is quenched from higher temperatures, a larger number of vacancies will be quenched in. If all of the vacancies, or even a constant fraction, are quenched in, H_v can be obtained from the variation of the quenched-in resistance with the quenching temperature T_q. The quenched-in concentration of vacancies will be orders of magnitude larger than the initial concentration, so that Eq. (2-40) can be written

$$\Delta \rho / \alpha = \Delta N_v = N_v = A \exp(-H_v(qu)/RT) \qquad (2\text{-}41)$$

This technique has been used by several groups, and consistent results have been obtained for many metals. Representative values are given in Table 2-1.

The temperature dependence of the rate of annealing out of vacancies from quenched samples allows one to measure H_m. However, it now is generally agreed that the most accurate values of H_m can be obtained by subtracting H_v from the activation energy for self diffusion in the temperature range where monovacancy diffusion is dominant, Q_1.

Positron Annihilation. When certain radioactive isotopes decay they emit particles called positrons. These have the mass of an electron but their charge is equal and opposite to that of the electron. They could also be called 'anti-electrons' since when a positron combines with an electron the mass of the two particles is converted into energy in the form of two gamma rays.

A schematic representation of positron annihilation is shown in Fig. 2-16. A source containing Na^{22}, often in the form of NaCl, is placed near the sample that is to be studied. When a Na^{22} atom decays it emits

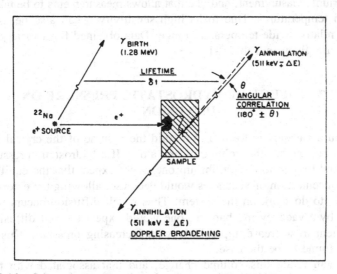

Fig. 2-16—Schematic representation of positron annihilation indicating the basis for the three experimental techniques: lifetime, Doppler broadening and angular correlation. [R. W. Siegel, *J. Nucl. Matl.* 69 (1978) 117.]

a 1.12 MeV γ ray and an energetic positron. When the positron enters the solid it quickly loses its kinetic energy and becomes thermalized. It migrates through the solid until after between zero and a few hundred pico-seconds it combines with an electron to form two gamma rays of about 511 keV. The great sensitivity of positrons to vacancies in metals (down to N_v of 10^{-6}) arises from the tendency of a positron to become trapped in a bound state in a vacancy where positron annihilation subsequently occurs. The absence of core electrons in a vacancy relative to the electron density in the perfect lattice, results in the lifetime of positrons localized in the vacancy trap being 20 to 80% longer than those in the perfect lattice.

If the temperature of the source and specimen are raised from a low temperature the vacancy concentration increases. With this rise in vacancy concentration the average lifetime of the positrons in the metal increases as more of them become trapped in vacancies and are annihilated there rather in the perfect regions of the lattice. Ultimately when essentially all of the positrons are trapped and annihilated at vacancies the mean lifetime no longer increases with temperature. The temperature dependence of N_v can also be obtained from related small variations in the relative energy or direction of the γ rays emitted on annihilation.

Positron trapping occurs in most, though not all, metals. It is an equilibrium measurement, and one that allows measurements to be taken at high temperatures. Due to its high sensitivity it can also be used over a relative wide temperature range. Data obtained for a variety of metals are given in Table 2-1.

2.7 EFFECT OF HYDROSTATIC PRESSURE ON DIFFUSION

When a vacancy is formed in a solid the volume of the crystal increases by roughly the volume of one atom. If a hydrostatic pressure is applied to a solid at equilibrium one might expect that the equilibrium concentration of vacancies would decrease, allowing the external pressure to do work on the system. Thus if self diffusion occurs primarily by a vacancy mechanism, one would expect the self diffusion coefficient to decrease appreciably with increasing pressure. This is indeed found to be the case.

One can relate this volume change, and that associated with the movement of atoms through the lattice, by applying equilibrium thermodynamic arguments. From thermodynamics we have the relation

$$(\partial G / \partial P)_T = V \qquad (2\text{-}42)$$

Considering G_m and G_v, to be free energies and using the equation for D

$$D = a_o^2 N_v w = a_o^2 \nu \exp(-G_v/RT) \exp(-G_m/RT) \qquad (2\text{-}43)$$

leads to the equation

$$\left[\frac{\partial \ln(D/a_o^2\nu)}{\partial P}\right]_T = \frac{-1}{RT}\left[\left(\frac{\partial G_v}{\partial P}\right)_T + \left(\frac{\partial G_m}{\partial P}\right)_T\right] \qquad (2\text{-}44)$$

and

$$\left[\frac{\partial \ln(D/a_o^2\nu)}{\partial P}\right]_T = \frac{-1}{RT}(V_v + V_m) = -(V_{SD}/RT) \qquad (2\text{-}45)$$

V_{SD} is called the activation volume. V_v is the partial molar volume of the vacancies. Its magnitude will depend on the degree to which the atoms surrounding a vacancy relax into it. If there was no such relaxation, V_v would equal the molar volume of the metal Ω since creating a mole of vacancies in the specimen would increase the volume of the large piece just as much as adding a mole of the pure metal would. However, there will be some relaxation of the lattice into the vacancy, so V_v will be less than Ω. For a close-packed metals like Cu, V_v should be an appreciable fraction of Ω. For bcc alkali metals like Na or Li it is a somewhat smaller fraction of Ω.

The second term contributing to the activation volume V_m is the partial molar volume of activated complexes. At the saddle point the diffusing atom is expanding a constriction, while the volume of the divided vacancy is approximately unchanged. Thus, V_m is probably positive, but small. This means that an increase in the pressure would decrease the concentration of activated complexes slightly.

Table 2-2 gives some experimental values of the ratio V_{SD}/Ω. It is positive and less than unity for all of the solids. Vacancy diffusion is the dominant mechanism here so this indicates that there is an appreciable relaxation of the lattice into a vacancy. Note that the value of V_{SD}/Ω for fcc metals is larger than that for the less closely packed lithium or sodium lattices. (The larger of the two values for Na is found at higher temperatures and is associated with divacancy diffusion. The smaller occurs at low temperature where the monovacancy mechanism dominates diffusion.) Experiments on the effect of pressure on the rate at which quenched-in vacancies anneal out show that increasing the pressure decreases the rate of annealing. In gold, V_m/Ω = 0.15.[19] This indicates that V_m is roughly one fourth that of V_v.

---

[19]R. Emrick, *Phys. Rev.*, 122 (1961) 1720.

Table 2-2. Activation Volumes for Self-Diffusion

Metal	V_{SD}/Ω	Ref.
Solid self-diffusion		
Ag	0.9	b
Au	0.72	b
Cu	0.9	b
Li	0.28	b
Na	0.32, 0.59	a
Pb	0.73	b
Liquid diffusion		
Hg	0.04	c
Ga	0.048	c
Interstitial		
C in Fe (250° K)	0.003	d
N in V (433° K)	0.01	d
Cu in Pb (600° K)	0.004	d

[a]J. N. Mundy, *Phys. Rev. B3* (1971) 2431.
[b]N. L. Peterson, *J. Nucl. Matl.*, *69* (1978) 3.
[c]N. Nachtrieb, *Liquid Metals and Solidification*, ASM, Metals Park, OH, (1958), p. 49.
[d]D. Beshers in *Diffusion*, ASM, Metals Park OH, (1973) p. 209.

It has been proposed that diffusion in a liquid occurs when a hole opens up in the nearest-neighbor shell of atoms, and the diffusing atom jumps into it. Some people conceive of these holes as having a volume comparable to that of a vacancy in a crystal. The effect of pressure on D is an ideal way to measure the mean size of these holes. The measured values of V_{SD} indicate that, at least in mercury and gallium, these holes are such a small fraction of Ω that the concept of vacancies in liquids is not appropriate.

Note that the activation volumes for interstitial diffusion are an order of magnitude lower than those for self diffusion by a vacancy mechanism. This stems largely from the fact that no vacancy need be formed for interstitial diffusion. Copper dissolves substitutionally in lead, but is believed to move by an interstitial mechanism. This will be discussed further in Chap. 3.

2.8 EMPIRICAL RULES FOR OBTAINING Q AND D_o

Just as there are systematic variations of physical properties with position in the periodic table, e.g. melting point, elastic constants, etc., there is a systematic variation of diffusion data. This provides a way to check new data for consistency, and can be used for making informed guesses about the value of D for self diffusion where no measurements are available. Brown and Ashby have assembled data for a wide variety of solids, chosen three correlations, and applied them to a wide range of types of solids.[20] Their equations are:

- The diffusion coefficient at the melting temperature $D(T_m)$ is a constant.
- The ratio of activation energy to RT is a constant, Q/RT_m.
- The activation volume is given by the equation

$$V^* = (Q/T_m)(dT_m/dp)$$

(The activation volume reflects the effect of pressure on D and is discussed in Sect. 2-9.) Some of their mean values for various crystal structures are given in Table 2-3.

Table 2-3. Diffusion Correlations

Class	$D(T_m)(cm^2/s)$	Q/RT_m	V^*/Ω
Fcc	5.5×10^{-9}	18.4	0.85
Bcc (Li,Na,K,Rb)	1.4×10^{-6}	14.7	0.40
Bcc (trans. elements)	2.9×10^{-8}	17.8	
Hcp (Mg,Cd,Zn)	1.6×10^{-8}	17.3	0.68
Alkali Halides	3.2×10^{-9}	22.7	4.6

From A. M. Brown, M. F. Ashby, *Acta Met.*, **28** (1980) 1085.

One of the better founded rules is that of Zener for the entropy term S_m. This was originally put forward as a theory when the available experimental values of D_o for interstitials varied by many orders of magnitudes and not even the sign of S_m could be inferred with any certainty. Zener reasoned that much of the work in moving an atom from an equilibrium position to the saddle point goes into elastically straining the lattice around the saddle point. Thus the work G_m can be set equal to a constant times $\lambda(\epsilon_o)^2$ where ϵ_o is some representative strain for the matrix when the atom is at the saddle point and where λ is an appropriate elastic modulus for the solvent. Since G_m is a Gibbs free energy, it follows that

$$S_m = -(\partial G_m/\partial T)_p \qquad (2\text{-}46)$$

But ϵ_o is essentially independent of temperature, so G_m varies with temperature in the same way as λ. Experimentally it is found that $d\lambda/dT$ is negative for all solids not undergoing a phase change; therefore, S_m will be positive. The vibration frequency of the interstitial ν can be estimated by assuming that the potential energy varies sinusoidally from

[20]See also, H. Bakker, *Diffusion in Crystalline Solids*, ed. G. E. Murch, A. S. Nowick, Academic Press, 1984, p. 189–258.

the equilibrium position to the saddle point. In this case

$$\nu = (Q/m\alpha^2)^{0.5} \tag{2-47}$$

where m is the mass of the interstitial, and α the jump distance. This immediately set a lower limit on D_o (interstitial) given by the equation

$$D_o > a_o^2 \nu \simeq 10^{-3} \ \text{cm}^2/\text{s} \tag{2-48}$$

Zener developed this argument and predicted that S_m is given by the equation

$$S_m \simeq \beta(Q/T_m) \tag{2-49}$$

where much of the variation of the elastic constants from one element to another can be removed by using the parameter.

$$\beta = -d(\mu/\mu_o)/d(T/T_m)$$

Here T_m is the melting point of the solvent, and μ_o is the elastic modulus at a low temperature.

Table 2-4 presents some of the accurate data now available. The prediction of Zener's theory can be represented as a constant value of

$$\lambda = S_m/(\beta Q/T_m)$$

where λ is the ratio of the experimental value of S_m to the value predicted by Eq. (2-49). The agreement may not appear to be outstanding,

Table 2-4.   Diffusion of Interstitials in Bcc Metals

Solvent	Solute	D_o (cm^2/s)	Q (kJ/mol)	S_m/R	λ
V	C	0.0088	116	1.4	0.66
	N	0.042	148	2.9	1.1
	O	0.025	123	2.5	1.2
Nb	C	0.010	141	1.3	0.43
	N	0.009	145	1.2	0.37
	O	0.021	113	2.3	0.92
Mo	C	0.012	161	1.5	0.61
	N	0.0043	109	0.74	0.44
	O	0.028	105	2.7	1.7
Fe	C	0.004	80.1	0.83	0.37
	N	0.005	76.8	1.1	0.51
	O[b]	0.002	86.0		

[a]From D. Beshers in *Diffusion*, American Soc. Metals, Metals Park, OH, 1973, pp. 209–40.
[b]J. Takada, *Oxidation of Metals*, 25 (1986) 93–103. (Internal oxid.)

and indeed it is not. However, the very accurate data which are now available have been almost entirely determined since 1950. The data available then were much poorer, and Zener's theory was instrumental in casting doubt on the validity of much of the older data.

Self diffusion data for several pure elements, as well as Q/T_m are given in Table 2-5.

2.9 DIVACANCY FORMATION

If two vacancies are on adjoining sites, they are said to form a divacancy. There is an attraction between vacancies in many metals and thus a tendency to form divacancies. In addition divacancies often move through the lattice more rapidly than do monovacancies. Thus divacancy formation can enhance the rate of self diffusion by a vacancy mechanism.

Consider the reaction of two vacancies to form a divacancy. If there is no interaction between the two, the atom fraction of such pairs N_v^2 in an fcc lattice is given by the equation

$$N_{v2} = 6(N_v)^2 \qquad (2\text{-}50)$$

This is obtained as follows. The number of vacancies that are in divacancies in a mole of material is given by the number of vacancies times the probability that there will be a vacancy on any one of the

Table 2-5. Self Diffusion in Metals (Constant Q)

Metal	Struct	D_o (cm^2/s)	Q (kJ/mol)	Q/T_m (kJ/K)
Cr	bcc	970	435	0.20
α-Fe	bcc	2.0	239	0.20
K	bcc	0.16	39.5	0.12
Mo	bcc	0.1	386	0.13
W	bcc	1.88	586	0.16
Al	fcc	0.047	123	0.13
Co	fcc	0.83	284	0.16
γ-Fe	fcc	0.49	284	0.17
Pb	fcc	1.37	109	0.18
Pd	fcc	0.21	266	0.15
Pt	fcc	0.33	285	0.14
Th	fcc	1.2	320	0.16
Ge[1]	dia.cub	25	318	0.26
Si[2]	dia.cub	20	424	0.25

Data from N. L. Peterson, *J. Nucl. Matl.*, *69&70* (1978) 3.
[1] G. Vogel, G. Hettich, H. Mehrer, *Jnl. Phys. C*, *16* (1983) 6197.
[2] F. J. Demond, et al., *Phys. Lett. 93A* (1983) 503.

vacancy's twelve nearest-neighbor sites. In the absence of any inter-action between vacancies, the probability that there is a vacancy on any specific site is N_v. If the number of vacancies is n_v then the number of vacancies in divacancies is $12\ n_v N_v$. This means that the number of divacancies is $6\ n_v N_v$, and if we define N_{v2} as this number over the number of sites in a mole, Eq. (2-50) is obtained.

If there is an interaction between two vacancies, a correction term must be added to Eq. (2-50). The more interesting situation is the one in which the two vacancies are attracted to one another. In this case, the free energy of the lattice is lowered when two vacancies move together to form a divacancy, or equivalently, when work is required to separate two adjacent vacancies. In this case N_{v2} will be greater than $6(N_v)^2$. The equation for N_{v2} can most simply be obtained by forming the equilibrium constant for the reaction between monovacancies to form a divacancy. If we consider the formation of divacancies of a particular orientation, the equation can be schematically written

$$\square + \square = \square\square$$

and, from the law of mass action[21]

$$ln(N_{v2}/N_v^2) = -G_b/RT \quad \text{or} \quad N_{v2} = N_v^2 \exp(-G_b/RT) \quad \text{(2-51)}$$

However, there are six different orientations of divacancies in an fcc lattice, and we wish to count all of them in N_v^2. Therefore,

$$N_{v2} = 6(N_v)^2 \exp(-G_b/RT) \quad \text{(2-52)}$$

Here G_b is the molar free-energy change of the lattice when divacan-cies are formed from two initially separated vacancies. If there is an attraction between vacancies, the energy is negative and one speaks of a binding energy between the vacancies in a divacancy.

The binding energy for divacancies is roughly equal to 10% of H_v. Thus though the exponential term in Eq. (2-52) gets larger as the tem-perature drops, N_{v2} at equilibrium will decrease much more rapidly with temperature because N_v^2 drops much more rapidly with tempera-ture. Thus the contribution of divacancies to self diffusion would be greatest at high temperatures. On the other hand if a metal is rapidly quenched from high temperatures the vacancy concentration doesn't change during the quench, the $\exp(-G_2/RT)$ term becomes larger, and divacancies can make a large contribution to the diffusion of excess (non-equilibrium) vacancies to sinks. For example the divacancy con-

---

[21]The reader who is not familiar with this law will find it discussed under this title or "equilibrium constant" in most books on thermodynamics or physical chemistry, e.g., L. Darken, R. Gurry, *Physical Chemistry of Metals*, chap. 9, McGraw-Hill Book Company, Inc., New York, 1953.

tribution can substantially enhance the rate of solute diffusion and the rate of precipitation in age hardening alloys.

2.10 SELF DIFFUSION ANOMALIES

It appears that self diffusion in pure metals always occurs by the vacancy mechanism. Often it occurs with an activation energy Q that is constant over a wide range of temperature and many orders of magnitude of D_{SD}. However, as more precise data have become available over a wider temperature range it has become clear that the Q is not always constant; that is, a plot of $ln\ D$ vs. $1/T$ is curved, not straight. This behavior is somewhat arbitrarily termed "anomalous." At least two mechanisms have been identified that can account for this, divacancy diffusion in addition to monovacancy diffusion, and the decrease of H_m as a phase transition is approached.

Divacancy Diffusion. It was pointed out in Sec. (2-8) that at higher temperatures the equilibrium concentration of divacancies can become appreciable, especially if there is an attraction between the vacancies. These divacancies diffuse faster than monovacancies, and independently of the monovacancy. That is they both contribute to Γ in the equation $D = \Gamma a^2/6$. In this case the self diffusion can be described by adding the contributions of the two mechanisms to give an equation of the form

$$D = D_{o1} \exp(-Q_1/RT) + D_{o2} \exp(-Q_2/RT) \qquad (2\text{-}53)$$

Here D_{o1} and Q_1 are those for the monovacancy that were treated above. The energy $Q_2 = H_m + 2H_v - H_b$, where H_b is the enthalpy part of G_b discussed in Sec. (2-8), and is positive if there is attraction between the vacancies. Table 2-6 gives data for several metals whose behavior is better described this way. Fig. 2-17 shows a plot of $ln\ D$ vs. $1/T$ for self diffusion in sodium. The curvature is clearly visible. Study of the effect of pressure on D indicates that the activation volume varies with pressure in a manner that supports the operation of a divacancy at high temperature and a monovacancy at lower temperatures.[22]

Phase Changes. Fig. 2-18 shows the variation of $ln\ D$ with $1/T$ for self diffusion of Zr^{95} in β-Zr. The curvature is continuous, pronounced, and far greater than is found in metals where there is no phase transition but only a divacancy contribution. In Fig. 2-18 the intercepts and slopes of the highest and lowest temperature tangents give, $4.8 \times 10^{-6} < D_o < 2.5 \times 10^{-2}$ cm²/s, and $86.6 < Q < 196$

[22]J. N. Mundy, *Phys. Rev. B3* (1971) 2431. See also N. Peterson, *J. Nucl. Matl.*, 69, (1978) 3.

Table 2-6.   Self Diffusion in Pure Metals (Variable Q)

Metal	Struct	$D_{\varrho 1}$ (cm²/s)	Q_1 (kJ/mol)	$D_{\varrho 2}$ (cm²/s)	Q_2 (kJ/mol)
Ag	fcc	0.04	170	4.7	211
Au	fcc	0.04	170	0.56	229
Cu	fcc	0.16	200	6.4	250
Ni	fcc	0.92	278	37	357
Li	bcc	0.038	50	9.5	67
Nb	bcc	0.008	349	3.7	438
Na	bcc	0.004	35.2	0.29	44
Ta	bcc	0.018	392	10	516
V	bcc	0.014	283	7.5	359
β-Ti	bcc	0.00036	130	1.1	251
β-Zr	bcc	0.000085	116	2.8	273

From N. L. Peterson, *J. Nucl. Matl.*, *69&70* (1978) 3–37.

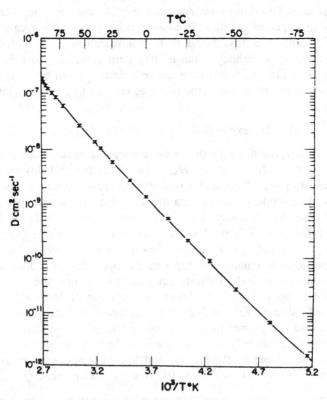

Fig. 2-17—Log D vs. $1/T$ for sodium self diffusion in sodium. [J. Mundy, *Phys. Rev.*, *B3* (1971) 2431.]

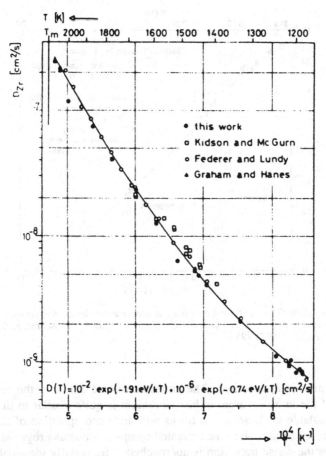

Fig. 2-18—Log D vs. $1/T$ for Zr self diffusion in Zr. [C. Herzig, H. Eckseier, *Zeit. Metal.*, 70 (1979) 215–23.]

kJ/mol)[23]. The bcc beta phase of zirconium is stable from the melting point down to 863° C. Below this temperature the beta transforms to the hcp alpha phase. One reason a phase becomes unstable on cooling is that the elastic constants resisting certain vibrational modes (shears) in the lattice become weak. As the shear waves in question become larger in amplitude the matrix phase becomes less stable relative to the new phase, and finally below a certain temperature a new phase becomes stable. Sanchez and de Fontaine have shown that in β-Zr this decrease in activation energy is closely related to the structural fluctuations that become increasingly frequent and large as the transition

[23]J. Federer, T. S. Lundy, *Trans, AIME*, 227, (1963) 592.

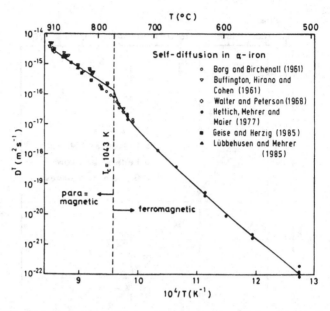

Fig. 2-19—Log D vs. $1/T$ for Fe in α-Fe at temperatures below the paramagnetic-ferromagnetic transition (Curie) temperature. [From G. Hettich, H. Mehrer, K. Maier, *Scripta Met.*, *11* (1977) 795.]

temperature is approached. The configuration of atoms at the saddle point of the jumping atom in bcc zirconium is quite similar to that of the metastable ω phase which forms with the decomposition of the β-Zr. Thus the formation of the activated complex becomes progressively easier as the phase transition is approached.[24] Essentially identical behavior is well documented for bcc β-Ti, and low values of D_o and Q have been reported for self diffusion in bcc Hf as well as the bcc phases in actinide and rare earth metals which undergo phase transitions. In going from β to α in Zr, the self diffusion coefficient drops by five orders of magnitude as D falls to values for fcc metals without a phase transition.

One final example of the effect of a phase transition is found for self diffusion of iron when it goes through the paramagnetic-ferromagnetic transition at 770° C. Fig. 2-19 shows a plot of *ln D* vs. $1/T$. Here there is no change in the crystal structure, but only a change in the electronic structure and magnetic properties.

[24]J. M. Sanchez, D. de Fontaine, *Acta Met.*, *26*, (1978) 1083.

PROBLEMS

2-1. In pure Ni it is believed that an interstitial Ni atom rests on the {100} plane, and shares one of the lattice sites with the atom originally there. In a sequence of drawings of a (100) plane show the atom movements involved in diffusion of the interstitial by an interstitialcy mechanism.

2-2. If at $t = 0$, a quantity of solute is located at the point $r = 0$ in a three dimensional medium, the concentration of solute at any point r from the origin, after time t, is

$$c(r,t) = (B/t^{1.5}) \exp(-r^2/4Dt)$$

(a) Give the probability (normalized to one) of finding an atom in a spherical shell dr thick and r from the origin.

(b) What is the mean-square value of r, r^2, for the solute after time t?

(c) Using the results of part (b) and the random-walk equations $\overline{r^2} = n\alpha^2$, show that $D = \Gamma\alpha^2/6$ when $\Gamma = n/t$. (Note this provides a rigorous derivation of Eq. 2-14).

2-3. An adsorbed W atom on an atomically smooth surface of W can be observed at low temperatures with the field ion microscope. If the sample is heated, then cooled again and observed, the atom has moved a distance Δx. Δx is measured after each of several anneals.

(a) On a plane (2-dimensional diffusion) the mean square distance an atom diffuses in time t is $4Dt$ rather than $6Dt$ as in 3-dimensions) Give an equation that can be used to calculate the surface diffusion coefficient for the adsorbed atom D_a.

(b) If the following values of Δx (in nm) are observed after repeated 100 s anneals (1.0, 0.9, 1.5, 0.5, 1.1), calculate D_a

2-4. (a) Calculate γ for a tracer in a pure bcc metal where γ is defined by the equation:

$$D = \gamma a^2 w N_v$$

(b) Calculate γ for an interstitial solute in a dilute bcc binary alloy.

2-5. In hydrogen gas at 1 atm and 25° C, the average molecular velocity is 1.3×10^5 cm/s, and the mean free path is 1.9×10^{-5} cm. Calculate the diffusion coefficient of the gas. (Take the average velocity to be the same as the root-mean-square velocity.)

2-6. In the temperature range -70 to 400° C the diffusion coefficient for C in α-Fe is $D = 0.02 \exp(-10120/T)$ cm^2/s. If the average vibration frequency of the carbon atom in the lattice is $5.0 \times$

10^{13}/s, calculate the entropy of motion, S_m, for carbon diffusion. (Take the lattice parameter to be 2.48×10^{-8} cm.)

2-7. Concerning the annealing out of excess vacancies in gold:

(a) If H_m of a vacancy is 79 kJ/mol, estimate D_v at 40° C.

(b) If $\tau = d^2/\pi^2 D_v$, estimate the mean intersink distance, d, if $\tau = 200$ hr at 40° C.

(c) The activation energy for the diffusion of a divacancy pair, H_{m2}, is 60 kJ/mol. Assuming that D_o is the same for both D_v and D_{v2} (divacancies), calculate the ratio Dv/D_{v2} at 40° C.

2-8. Would you expect the activation volume for self diffusion by an interstitialcy mechanism to be larger or smaller than for a vacancy mechanism? Explain.

Answers to Selected Problems

2-2. (b) $\overline{r^2} = 6 Dt$

2-3. (b) $D = 2.75 \times 10^{-17}$ cm^2/s

2-4. (a) $\gamma = 1$, (b) $\gamma = 1/6$

2-5. 0.41 cm^2/s

2.6. $S_m = 2.7$ cal/mol

2-7. (a) $D_o(SD) = D_{ov}\exp(S_v/R) = 0.09$ cm^2/s. If $S_v/R = 1$, $D_v = 1.6 \times 10^{-14}$ cm^2/s.

(b) $d = 1.1$ μm. (c) $D_v/D_{v2} = 6.7 \times 10^{-4}$.

3
DIFFUSION IN DILUTE ALLOY

The next degree of complexity after studying diffusion in pure metals is to study the diffusion in dilute alloys. The simplest problem in this area arises in interstitial alloys. Here the solute atoms diffuse on a sublattice whose sites are essentially all vacant, and the only role played by the solvent atoms is to form the barriers which define the sublattice of the interstitial sites. Because the two types of atoms do not share the same sites, the theory of interstitial diffusion is relatively simple and has been discussed in Chap. 2. The use of relaxation or resonance techniques to measure D for interstitials in bcc metals is introduced as a representative of a family of techniques in which the mean jump frequency of the interstitials is obtained from some relaxation phenomenon. This frequency is then combined with a model and random-walk theory to give values of D.

The problem of an atomic theory of D for substitutional solute in a metal is different. Here the solute and the solvent atoms share the same sites, and the analysis of D becomes more complex. Since the two atoms do share the same sites, the difference in D for the solute and solvent atoms allows one to estimate the ratio of the jump frequencies for the solute and the solvent atoms. There are several theories which suggest why, and by how much these D's should differ. The data on most substitutional alloys fit these theories, but some don't and these anomalies are discussed briefly.

Finally, the matter of trapping is discussed. There is often an attraction between solutes and structural inhomogeneities in the solid, in effect a tendency to segregate there. This can markedly influence the rate at which a solute can diffuse out of, or through, a solid. The most striking effects are seen for hydrogen where the trapping energies can be large relative to the activation energy for diffusion in the absence of traps. As a result the apparent diffusion rate of hydrogen in a metal like iron can be determined entirely by the density of traps rather than by the rate of lattice diffusion.

3.1 INTERSTITIALS AND ANELASTICITY[1]

In a bcc lattice such as alpha-iron, an interstitial atom such as carbon strains the lattice more along one of the cubic directions of the lattice than along the other two. If the interstitial jumps to a neighboring site, the direction of this high strain changes. If a stress is applied to the crystal, the energy of the strain energy of the lattice will be lowered if the interstitials jump to sites which align their strain fields with that of the applied strain. This alignment gives rise to an additional strain called the anelastic strain. From the rate at which this anelastic strain appears, the jump frequency can be determined and from this the diffusion coefficient.

To give a clearer picture of the distortion associated with an interstitial solute, consider the interstitial atom shown as a solid black circle in the bcc lattice of Fig. 3-1. Its two nearest neighbors are shown as circles con taining crosses, and their normal sites are a distance of $a_o/2$ from the center of the solute. The four solvent atoms which are next nearest neighbors to the interstitial (labeled e, f, g, and h) lie in the xy-plane; their centers are $a_o/\sqrt{2} = 0.71\ a_o$ from the center of the solute atom. If now the matrix atoms in Fig. 3-1a are enlarged until they touch one another, as is a reasonable model for a transition metal, it is seen that the distortion caused by the interstitial atom shown will be much more severe in the z direction than in either the x or the y direction. Thus the strain field introduced by the solute is said to have tetragonal symmetry.

If the interstitial now jumps to the interstitial site to its left (shown by a small black circle in Fig. 3-1), its surroundings will be equivalent, but the tetragonal axis of the distortion will be in the y direction. If the site occupied by the interstitial in Fig. 3-1 is called a z site, consideration of Figs. 3-1a and 3-1b will show that atoms on z sites have only x and y sites as nearest neighbors; similarly, x sites have only y and z neighbors, etc.

Consider now a bcc single-crystal wire with the [001] direction along its axis and its interstitial atoms uniformly distributed between x, y, and z sites. If a small weight is hung on the wire, there will be an

---

[1]For a more complete treatment see A. S. Nowick, B. S. Berry, *Anelastic Relaxation in Crystalline Solids*, (Academic Press, New York 1972).

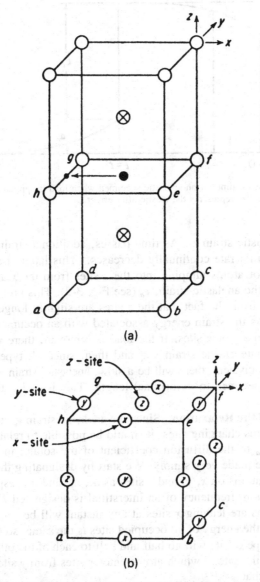

(a)

(b)

Fig. 3-1—Body-centered cubic lattice showing interstitial atom (●), its two nearest neighbors (⊗), and an arrow pointing to one of the four sites the interstitial can jump into. (b) shows the interstitial sites in a bcc unit cell. There are no sites inside the unit cell so all of the types of sites are shown on the visible surfaces of the cube.

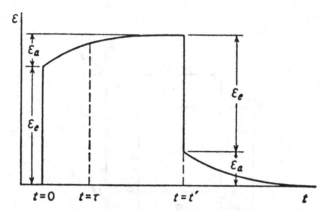

Fig. 3-2—Strain vs. time for an anelastic specimen when load is applied at $t = 0$ and removed at $t = t'$. ϵ_a represents the elastic after effect.

immediate elastic strain ϵ_e. As time passes, additional strain appears, though the strain rate continually decreases.[2] This latter strain is due to a net flux of atoms jumping into the z sites from the x and y sites and is called the anelastic strain, ϵ_a (see Fig. 3-2). This preference for z sites comes from the fact that the z sites are slightly longer in the z direction. Thus the strain energy associated with an occupied z site is less than for the x or y sites. If the load is removed, there will again be an immediate elastic strain $-\epsilon_e$ and then, since all types of sites will now be equivalent, there will be a slow, anelastic[3] strain $-\epsilon_a$ which restores the specimen to its original length. This is called the elastic after effect.

Analysis of the Relaxation. Since the anelastic strain ϵ_a stems from interstitial atoms changing sites, it should be possible to relate the rate of decay of ϵ_a to the diffusion coefficient of the solute; in fact, this relation can be made very simply. We start by designating the number of interstitial atoms on x, y, and z sites as n_x, n_y, and n_z, respectively. If the mean jump frequency of an interstitial is designated Γ, the rate at which atoms are leaving x sites at any instant will be Γn_x. At zero applied stress the energy of all occupied sites is the same, so the atoms leaving one type of site will go half and half to each of the other types. For example, the rate at which atoms enter x sites from y sites is $\Gamma n_y/2$. It follows that

---

[2]The weights involved are such that the elastic strain is low, 10^{-5} or less.

[3]Since the strain is not a single-valued function of the stress, the wire is not elastic. However, when the stress is removed, the strain returns to zero. To differentiate this type of nonelastic behavior from the type in which a permanent set occurs, Zener coined the word "anelastic."

$$\frac{dn_x}{dt} = -\Gamma n_x + \frac{\Gamma n_y}{2} + \frac{\Gamma n_z}{2} \qquad (3\text{-}1)$$

but since $n_x + n_y + n_z = n$ is a constant, we can replace $n_y + n_z$ to give

$$\frac{dn_x}{dt} = \frac{d(n_x - n/3)}{dt} = -\frac{3}{2}\left(n_x - \frac{n}{3}\right)\Gamma \equiv -\frac{3}{2}\Delta n \Gamma \qquad (3\text{-}2)$$

The literature on this subject invariably talks in terms of the mean time of stay τ_j of an interstitial, which is $1/\Gamma$. Replacing $1/\Gamma$ by τ_j, Eq. (3-2) can be rewritten

$$d\Delta n/dt = -(3\Delta n/2\tau_j) = -\Delta n/\tau \qquad (3\text{-}4)$$

Here $2\tau_j/3$ has been replaced by the experimentally observed relaxation time τ. Integration gives

$$\Delta n = \Delta n_o \exp(-t/\tau)$$

Now ϵ_a is proportional to Δn, so if log ϵ_a is plotted versus t, τ can be obtained from the slope.

The random-walk equations of Sec. 2-3 enable us to relate τ to D. Equation (2-14) gave

$$D = (1/6)\Gamma \alpha^2 \qquad (2\text{-}14)$$

where α is the jump distance of the interstitial. Since $\alpha = a_o/2$ in a bcc lattice,

$$D = \Gamma a_o^2/24 = a_o^2/36\tau \qquad (3\text{-}5)$$

To measure ϵ_a and τ experimentally, it is customary to obtain the strain by twisting a thin wire instead of pulling it in tension. The advantage of using torsion is that, by attaching a mirror to the wire and using it to cast a reflected beam of light across the room, very small strains can be easily measured. A value of τ of almost an hour can be measured in this way. This corresponds to values of the diffusion coefficient of 10^{-20} cm^2/s whereas D is over 10^{-6} at the alpha-gamma transformation temperature of $910°$ C. These values of D are much smaller than those measurable by the thin-film tracer method since those methods require measurable penetration while the relaxation method reduces the required penetration to an absolute minimum: one atomic jump.

Resonance Techniques. Let us now see how this anelastic effect can be used at higher frequencies. Inspection of Fig. 3-2 shows that, after the stress is increased to some new, constant value, the elastic modulus (the ratio σ/ϵ) will be a function of time for $t < 2\tau$. If an oscillating stress is applied with a frequency $\omega \simeq 1/\tau$, this variation in the modulus, or relaxation, gives rise to a hysteresis loop when stress is plotted against strain. This hysteresis reflects a loss of energy, and is also called internal friction. If the stress is oscillated the maximum strain will occur shortly after the maximum value of the stress. Mathematically this lag of the strain behind the stress by a small angle δ is given by the equation[4]

$$\sigma = \sigma_o \sin \omega t \quad \epsilon = \epsilon_o \sin(\omega t - \delta)$$

When the stress and strain are out of phase in this manner, an energy ΔE is absorbed during each cycle. That ΔE is indeed nonzero when $\delta > 0$ can be seen by integrating over one cycle

$$\Delta E = \oint \sigma \, d\epsilon = \sigma_o \epsilon_o \int_o^{2\pi/\omega} \sin(\omega t) \cos(\omega t - \delta) \omega \, dt \quad (3\text{-}6)$$

Substituting $\cos(\omega t - \delta) = \cos \omega t \cos\delta + \sin \omega t \sin \delta$ and integrating gives

$$\Delta E = \epsilon_o \pi \sigma_o \sin \delta \qquad (3\text{-}7)$$

Thus when $\delta = 0$, there is no energy loss, and as δ increases from zero, so does ΔE.[5]

To give a qualitative feeling for the variation of δ with frequency, we refer to Fig. 3-3. Here stress is plotted vs. strain for three ranges of $\omega\tau$.

1. $\omega\tau \gg 1$. Here the applied frequency is so high that essentially no interstitials change sites in reaction to the applied stress. In this case, stress and strain are completely in phase; $\delta = 0$ and so $\Delta E = 0$.

2. $\omega\tau \simeq 1$. At this frequency, many of the interstitials will be able to change sites in reaction to the applied stress, but the stress will vary too rapidly for the equilibrium population of each type of site to be attained at any particular value of the stress. Thus $\delta > 0$, $\Delta E > 0$, and the lag between stress and strain shows up as a hysteresis loop.

3. $\omega\tau \ll 1$. Here the applied frequency is so much lower than $1/\tau$ that the population of each type of site will continually be in equilibrium with the applied stress. As a result, $\delta = 0$. The only difference between this case and that of $\omega\tau \gg 1$ is a smaller slope for the

---

[4]In systems of interest δ usually varies between zero and 0.1 radian.

[5]This energy loss is expressed in several different ways in the literature. If it is observed that δ is small, and noting that $(1/2)\epsilon_o \tau_o = E$, the most common expressions are: $\sin \delta = \tan \delta = \delta = \Delta E/2\pi E = Q^{-1}$

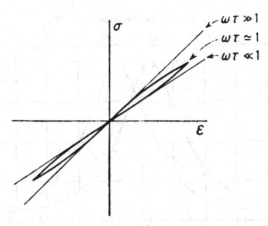

Fig. 3-3 — Stress-strain curves for very small strains in an anelastic material for various values of the product $\omega\tau$.

stress-strain line. This stems from the fact that here the strain at any stress is the elastic plus the anelastic strain, while with $\omega\tau \gg 1$ there is no anelastic strain.

A detailed analysis of this type of phenomenon will not be given. A plot of the observed hysteresis loss Q^{-1} ($=\delta$) versus $1/T$ is shown in Fig. 3-4. Since τ increases as $\exp(Q/RT)$, this is equivalent to a plot of δ vs $ln(\omega\tau)$. The curve agrees with the qualitative conclusions given above. A detailed analysis shows that the maximum in δ occurs at $\omega\tau = 1$.

In experimental studies of δ versus $\omega\tau$ there are a variety of techniques that can be used to drive the specimen, depending on the frequency needed.[6] In diffusion studies τ decreases exponentially with increasing temperature (since $\tau = 2/3\Gamma$), and it is easier to vary $\omega\tau$ continuously by changing the temperature than it is to vary ω. To determine τ, the system is oscillated at a fixed frequency, and the energy loss per cycle measured over a range of temperatures, allowing the curve of Fig. 3-4 to be traced. Because $\omega\tau = 1$ at the maximum in the curve, τ at the temperature of the maximum is $1/\omega$, that is $1/2\pi f$ where f is the frequency in cycles per second. If the applied frequency is increased and δ is again measured over the same temperature range, the curve will be shifted to higher temperatures, and ω can be determined at a second temperature. Once ω is known at two or more temperatures, ΔH can be calculated.

---

[6]For a review of techniques see D. Beshers, *Diffusion*, ASM, Metals Park, OH, (1973) p. 209.

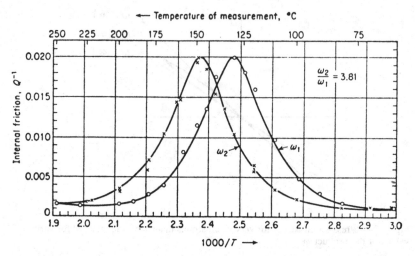

Fig. 3-4—Internal friction (δ) versus $1/T$ at two frequencies, for carbon in tantalum. [T. Ke, *Phys. Rev.*, *74* (1948) 9.]

A plot of data obtained for nitrogen in niobium is shown in Fig. 3-5. The values of τ between 20 and 6,000 s were obtained by measuring the elastic after effect while the values between 0.1 and 2 s were obtained from the rate of decay of the oscillations of a torsion pendulum. The two techniques combine to give values of τ (and thus D) ranging over five orders of magnitude. This wide range of values combined with the accuracy of the individual points give values of D_o and Q of high precision.

Two variants of this technique will be mentioned in closing. It was mentioned in Sec. 2-8 that vacancies in gold associate to form divacancies which can diffuse faster than individual vacancies. A similar association occurs between interstitial atoms. Here again the individual and paired interstitials give rise to separable peaks which allow the determination of the jump frequency of the pairs as well as the isolated interstitial. The magnitude of δ at its maximum is proportional to the number of atoms, or pairs which give rise to the peak in question. Thus the variation of the magnitude of δ at the peak due to pairs gives the variation in the concentration of such pairs.

3.2 IMPURITY DIFFUSION IN PURE METALS

In dilute substitutional alloys the solute and solvent share the same lattice sites and the same vacancies. (This is in contrast with the situation in ionic compounds like NaCl where the sodium and the chlo-

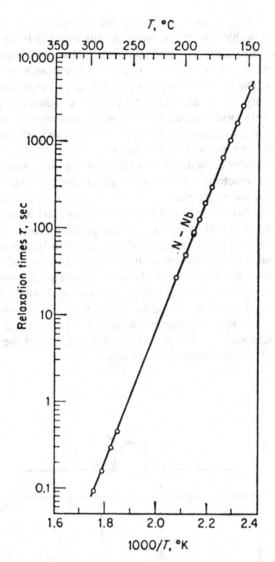

Fig. 3-5—Relaxation time vs. $1/T$ for nitrogen in Nb. [Courtesy R. Powers]

ride ions each diffuse on their own sublattice with their own point defects.) Two types in questions will be considered, what determines the relative mobilities of types of atoms in the alloy, and what is the relative effectiveness of a jump in giving rise to diffusion.

In Sec. 2-3 it was shown that when a vacancy mechanism operates in a pure metal, the self-diffusion coefficient is determined by the frequency with which an atom will jump into a vacant neighboring site w, and by the probability that a given neighboring site is vacant p_v. In an infinitely dilute alloy, the problem is then to estimate whether, and by how much, w and p_v for a solute atom differ from w and p_v for a solvent atom. The treatments of this problem are all approximate. The most satisfactory theory deals with the electronic, almost electrostatic, interaction. This model works best for silver and copper base alloys. Consider the case of a solvent of valence $+1$ and an impurity with the same size ion core but with a valence of $+2$, for example cadmium in silver. The metal is modeled as a lattice of Ag^+ ions surrounded by a "gas" of free electrons. If one of the silver atoms is removed and a cadmium atom put in its place, a Cd^{++} ion forms on the lattice site, and contributes two electrons to the electron gas. The insertion of the cadmium ion gives a sharp change in the positive charge density ρ^+ (see Fig. 3-6a). There is an electrostatic attraction between the cadmium ion and the extra electrons it contributed to the electron gas. Fig. 3-6a shows the variation of the positive charge density and the electron charge density ρ_e around the cadmium ion. Loosely speaking, the ion's second electron is weakly bound to it and spends its time within one to two atomic diameters of the Cd^{++} ion. The Cd^{++} is then said to be "screened." The electrostatic potential around the screened impurity can be approximated by the equation

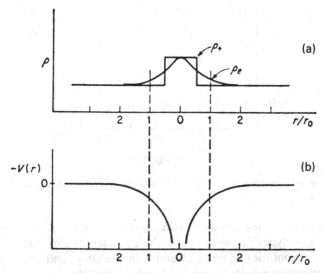

Fig. 3-6 — (a) Positive and negative charge density around a nucleus with a +2 charge in an array of nuclei with a +1 charge. r_o is the nearest-neighbor distance. (b) shows the resulting electrostatic potential.

$$V(r) = Ze/r \exp(-qr) \qquad (3\text{-}8)$$

where e is the charge on an electron, Z is the number of excess electrons per ion (one in the case of Cd in Ag) and r is the radial distance from the impurity. q is called the screening parameter and can be calculated from the free-electron theory of a metal. Figure 3-6b shows $-V(r)$ around the screened impurity below the assumed positive charge density distributions. $V(r)$ rises [$-V(r)$ falls] in the region one interatomic distance r_o from the ion. This indicates that a negative charge would be attracted by the Cd^{++} ion in this region. Since there is a net negative charge associated with a vacancy, the energy of a vacancy on a site next to a Cd^{++} ion will be reduced by an amount we will designate $E(r_o)$. This reduced energy of a vacancy next to a Cd^{++} ion, results in the concentration of vacancies on any given nearest-neighbor site being increased to

$$p_v = N_v \exp[E(r_o)/kT] \qquad (3\text{-}9)$$

In addition to making $p_v > N_v$ on a site next to a Cd^{++} atom the potential $V(r_o)$ makes H_m for the solute less than H_m for the solvent. The electrostatic attraction between the Cd^{++} and a vacancy tends to draw the vacancy and the Cd^{++} ion together and thereby reduce H_m.

The differences between Q for the solvent and Q for the solute, ΔQ, from this theory are compared with experimental results in Table 3-1. The agreement is quite good. Attempts to use the same theory for Cd and Zn base alloys are satisfactory, but the model does not fit data for Mg, Al, or transition metals as a solvent.

3.3 CORRELATION EFFECTS

In all cases studied up to this point the directions of successive jumps of the diffusing atom have been assumed to be independent of one another, that is they are uncorrelated. Also, the mean frequency of all

Table 3-1. Effect of Valence on Q (in kJ/mol)

	Copper Solvent				Silver Solvent		
Metal	Z	ΔQ_{ex}	ΔQ_{th}	Metal	Z	ΔQ_{ex}	ΔQ_{th}
Zn	+1	−20.5	−16.6	Pd	−1	+52.7	+30.9
Ga	+2	−19.2	−19.2	Cd	+1	−10.7	−10.2
Ge	+3	−24.0	−24.3	In	+2	−15.2	−14.2
As	+4	−35.5	−34.0	Sn	+3	−20.8	−19.4
Sn	+3	−23.2	−21.3	Sb	+4	−24.9	−23.8

Taken from A. D. LeClaire, *J. Nucl. Matl.*, *69&70* (1978) p. 82.

jumps has been assumed to be the same. In dilute substitutional alloys this is not true for either the solute or the solvent.

This section starts with a detailed discussion of the correlation between the successive jump directions of a radioactive isotope of A in pure A. In this manner the existence and effect of correlation between successive jumps is established. Following this, the effect of correlation on the equation for the diffusion coefficient of a substitutional impurity is given and discussed. The correction for correlation in pure metal will be seen to be small, but in the case of impurity diffusion, it can be quite pronounced.

Pure Metals. The successive jump directions of an atom will be uncorrelated if after any given jump all possible directions for the next jump are equally probable. For example, the jumps of a vacancy diffusing in a pure metal will be uncorrelated since after any jump all of the neighbors of the vacancy are identical. It follows that all possible jump directions have the same probability of occurring. This will not be the case for a tracer of A diffusing by a vacancy mechanism in pure A. After any jump of the tracer, all of its neighbors are not identical; one of them is a vacancy, and the most probable next jump direction for the tracer is right back to the site that is now vacant. This can be seen in the two-dimensional close-packed lattice in Fig. 3-7. If the tracer in the figure (at site 7) has just exchanged sites with the vacancy now at the site labeled 6, the tracer's most probable next jump is to return to 6. Its next most probably jumps are to sites 1 and 5; i.e., the vacancy jumps from 6 to 5 (Or 1) and then to 7. The tracer's least probably next jump is to 3, since this requires that the vacancy move from 6 around to 3 before it jumps to 7.

It was shown in Chap. 2 that if successive jumps are random then $\overline{R_n^2} = n\alpha^2$. However, if the direction of successive jump vectors are correlated then R_n^2 will be less than $n\alpha^2$. The degree of this reduction is corrected for with the correlation factor f which is defined as.

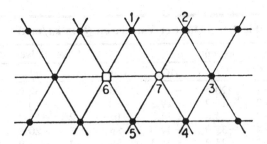

Fig. 3-7 — Portion of two dimensional close-packed lattice showing a tracer (○) and a vacancy (□).

$$f - \lim_{n \to \infty} (\overline{R_n^2}/n\alpha^2) \quad \text{or} \quad \overline{R_n^2} = n\alpha^2 f \qquad (3\text{-}10)$$

The equation for the tracer diffusion coefficient thus becomes $D = f\Gamma\alpha^2/6$ instead of $D = \Gamma\alpha^2/6$.

In Sec. 2-3 the equation derived for the mean value of R_n^2 was

$$\overline{R_n^2} = n\alpha^2 \left[1 + (2/n) \overline{\sum_j \sum_i \cos \theta_{i,i+j}} \right] \qquad (2\text{-}8)$$

Comparison of this with Eq. (3-10) shows that f equals the terms in square brackets. Thus the calculation of f requires that the mean value of $\cos\theta_{i,i+j}$ be determined. As was pointed out in Sec. 2-3, in a bcc or fcc lattice all jumps are equivalent except for their orientation. That is, all tracer-vacancy pairs that have just completed an exchange are indistinguishable, aside from their orientation. Thus the mean value of $\cos \theta_{i,i+j}$ is the same for each value of i, so it can be replaced by the mean value of the cosine of the angle between any jump and the subsequent jumps of the tracer, $\overline{\cos \theta_1}$, $\overline{\cos \theta_2}$, etc. The term in $\overline{\cos \theta_2}$ takes account of the correlation between the ith and $(i + 2)$th jump of the solute. Compaan and Haven show that $\overline{\cos \theta_i} = (\overline{\cos \theta_1})^i$ for a vacancy mechanism,[7] and the equation used to calculate f becomes

$$f = 1 + 2(\overline{\cos \theta_1})^1 + 2(\overline{\cos \theta_1})^2 + 2(\overline{\cos \theta_1})^3 + \ldots$$

$$= (1 + \overline{\cos \theta_1})/(1 - \overline{\cos \theta_1}) \qquad (3\text{-}11)$$

The final equation for f is the sum of the infinite series, as can be shown by dividing $1 + \overline{\cos \theta_1}$ by $1 - \overline{\cos \theta_1}$.

As an example of the calculation of f, we shall outline the evaluation of f for the two-dimensional close-packed lattice shown in Fig. 3-7. This requires the evaluation of $\overline{\cos \theta_1}$, the average value of the cosine of the angle between the last tracer jump (from 6 to 7) and the next tracer jump vector. For this it is necessary to calculate the probability of the tracer making its next jump to each of its six nearest neighbors. In general we define p_k as the probability that the tracer will make its next jump to its kth nearest neighbor. Similarly θ_k is the angle between the jump vector $6 \to 7$ and the jump vector $7 \to k$. The equation for $\overline{\cos \theta_1}$ is then

$$\overline{\cos \theta_1} = p_6 \cos \theta_6 + p_5 \cos \theta_5 + \ldots + p_1 \cos \theta_1 = \sum_{k=1}^{z} p_k \cos_k \qquad (3\text{-}12)$$

The basic problem here (as well as in the case of impurities) is to calculate the various p_k. The value of p_k is calculated by summing the

---

[7]See K. Compaan, Y. Haven, *Trans. Faraday Soc.*, 52 (1956) 786.

probability of the various vacancy trajectories which will move the tracer to site k on the tracer's first jump. Consider the case of the site labeled 6, that is, $k = 6$, p_6 can be calculated from the series

$$p_6 = n_{16} P_{16} + n_{26} P_{26} + \ldots = \sum_{i=1}^{m} n_{i6} P_{i6} \qquad (3\text{-}13)$$

where P_{i6} is the probability that the vacancy will return for the first time to site 7 from site 6 on its ith jump. This then moves the tracer to 6 on its first jump. n_{i6} is the number of paths which will allow the vacancy to move the solute from 7 to 6 for the first time on its ith jump. m is a large number fixed by the accuracy desired. Throughout this discussion, only one vacancy will be considered. Or equivalently, it is assumed that the density of vacancies is so low that no other vacancy will exchange with the tracer before the vacancy has randomized its position with respect to the initial solute-vacancy exchange.

Since the vacancy jumps are random, the probabilities of any particular vacancy path P_{i6} can be easily calculated. The probability that the vacancy will jump to any particular neighboring site on its first jump is $1/z$, or in this case $1/6$. The a priori probability that it will jump to one specified site and then to another specified site is $(1/6)^2$. In general then $P_{i6} = (1/6)^i$.

As the simplest possible case consider the value of $\cos \theta_1$ for the trajectories only one jump long, that is the vacancy exchanges with the tracer on its next jump ($m = 1$ in Eq. (3-13)). The probability of the vacancy exchanging with the tracer on its next jump is $1/z$. For this simplest case $p_1 = 1/z$ and $\cos \theta_1 = -1$. The other p_k equal zero at this level of approximation so $\overline{\cos \theta_1} = -1/z$ and

$$f = 1 - 2/z \qquad (3\text{-}14)$$

This is a good first approximation as can be seen in Table 3-2 which gives the exact values for several lattices as well as the value of the simple approximation $f = 1 - 2/z$. Note that this simple approximation accounts for about 90% of the correlation effect even though the jump of the tracer to only one of z neighboring sites is considered.

To pursue the problem somewhat farther, consider trajectories up to $m = 4$ in Eq. (3-13). There is just one path for the vacancy which will bring it back to site 7 on its first jump so $n_{16} = 1$. There are no paths which will return the vacancy to 7 from 6 on its second jump so $n_{26} = 0$. The value of n_{36} is five since a first jump by the vacancy to any of its nearest neighbors, aside from the tracer, and a return to 6 on the next jump will allow the vacancy to go to 7 from 6 on its third jump.

Table 3-2.   Values of f for Vacancy Diffusion in Various Lattices

	z	f	$1-2/z$
Two-dimensional:			
Square	4	0.46705	0.500
Hexagonal	6	0.56006	0.667
Three dimensional:			
Diamond	4	1/2	1/2
Simple cubic	6	0.65549	0.667
Body-ctr-cubic	8	0.72149	0.750
Face-ctr-cubic	12	0.78145	0.833

From: K. Compaan, Y. Haven, *Trans. Faraday Soc.*, 52 (1956) 786.

A similar examination shows that $n_{46} = 8$. The next level of approximation then gives

$$p_6 = 1/6 + 0 + 5(1/6)^3 + 8(1/6)^4 = 0.1960$$

By symmetry p_5 must equal p_1. To calculate these we note that $n_{15} = n_{11} = 0$, since the vacancy cannot return to 7 from 1 on its first jump. The equation for p_5 or p_1 carried out to the fourth vacancy jump is

$$p_5 = p_1 = 0 + (1/6)^2 + (1/6)^3 + 11(1/6)^4 = 0.0409$$

Similarly

$$p_4 = p_2 = 0 + 0 + (1/6)^3 + 2(1/6)^4 = 0.0062$$

$$p_3 = 2(1/6)^4 = 0.0015$$

Comparison of p_6 and p_3 shows that the tracer is 100 times more likely to make its next jump back to 6 instead of on to 3. Using Eq. (3-12)

$$\overline{\cos \theta_1} = (-1)0.196 + (-1/2)(2)0.0409$$

$$+ (1/2)(2)0.0061 + (1)0.0015 = -.2262$$

Eq. (3-11) gives $f = 0.631$. This is to be compared with the exact value $f = 0.560$ in Table 3-1. The difference between the two stems entirely from the omission of vacancy path involving $i > 4$. The fraction of the possible vacancy trajectories which has been omitted can be seen by noting that the sum of all return probabilities, p_1 through p_6, is only 0.292 instead of unity. Since the probability is unity that the tracer will make a next jump by exchanging with the vacancy initially at site 6, this means that the sum of p_k would increase over 70% if terms in higher values of i were included in the calculation of each

p_k.[8] Although 70% of the returns is a relatively large omission, it is seen that the value of f calculated here would take care of 89% of the total correlation. The main conclusion to be drawn is that the sums of p_k and f converge slowly to their final values. Hastening this convergence is the main problem in calculating f.

Very Dilute Alloys. In alloys, vacancies interact with impurities, so not only are the successive jumps of the impurities correlated, but the successive jumps of the vacancies are too. As an extreme example, consider the diffusion of an impurity which is "bound" or strongly attracted to a vacancy. To simplify the geometry, we again consider a two-dimensional close-packed lattice. Assume that the vacancy-impurity exchange rate w_2 is much greater than the vacancy-solvent exchange rate w_1.[9] Under these conditions the successive jump directions of the vacancy are no longer random but will be almost completely correlated. That is, if the vacancy exchanges with the impurity on a given jump, the probability that it will reverse that jump on its next exchange is almost unity. The angle between successive jumps would be π, and $\overline{\cos\pi} = -1$. Substituting this into equation (3-11) gives f = 0 as it should since for this type of exchange the net distance traveled is zero. In this case the impurity will translate through the lattice only as fast as the vacancy exchanges with the solvent atoms. Under these conditions the impurity diffusion coefficient will be given by the equation

$$D_2 = fa_o^2 w_2 p_v \simeq a_o^2 w_1 p_v \ll a_o^2 w_2 p_v \qquad (3\text{-}13)$$

Here $f \ll 1$, and it is obvious that correlation effects play a dominant role in determining D_2.

An approximate equation for f can easily be derived with the equations of the preceding subsection. First assume that the vacancy-impurity pair will not dissociate and that for its last jump the impurity exchanged with the vacancy shown in Fig. 3-8. The calculation of $\cos\theta_1$ is parallel to the case in which the vacancy jumps were random, except that in place of $1/z$ the probability that the vacancy and the impurity will exchange again on the next vacancy jump is $w_2/(w_2 + $

---

[8]That the sum of p_k will in fact equal 1 after a finite number of vacancy jumps can be shown by appealing to the continuum treatment of Chap. 1. If an instantaneous point source is placed in a two-dimensional medium at $t = 0$, the solution to the diffusion equation is $c(r,t)=(\text{const.}/t)\exp(-r^2/Dt)$. Thus the probability of finding a vacancy at its initial site in some time increment dt after time t is $p(t)dt=(\text{const}/t)dt$. Now the probability that the vacancy will return to its initial site at some time between $t = \epsilon$ and $t = t$ is the integral of $p(t)dt$ or proportional to $\ln(t)$, which must equal unity at some finite time. Therefore the vacancy will return to the tracer after some finite time.

[9]Here and in what follows the subscript 1 will refer to the solvent atoms while the subscript 2 will refer to the impurity.

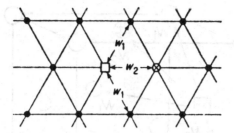

Fig. 3-8 — Two-dimensional close-packed lattice showing vacancy (□), impurity (⊗), and solvent atoms (●). This differs from Fig. 3-7 in that w_2 and w_1 are different.

$2w_1$).[10] If only the vacancy-impurity exchanges w_2 and w_1 are considered in calculating $\overline{\cos \theta_1}$, we obtain $p_6 = w_2/(w_2 + 2w_1)$. The approximate equations are

$$\overline{\cos \theta_1} = -w_2/(w_2 + 2w_1) \qquad (3\text{-}14)$$

and

$$f = w_1/(w_1 + w_2) \qquad (3\text{-}15)$$

Ignoring geometric constants, the equation for the diffusion coefficient of the solute atom is

$$D_2 = a_o^2 w_2 w_1 p_v (w_1 + w_2) \qquad (3\text{-}16)$$

Although this equation is only a first approximation, it has all of the characteristics of the exact solution for two (or three) dimensions. Inspection shows that D_2 is determined primarily by the slower of the two jump frequencies. If the impurity-vacancy exchange is much faster ($w_2 \gg w_1$), then $D_2 \simeq a_o^2 w_1 p_v$. If the reverse is true ($w_2 \ll w_1$), then $D_2 \simeq a_o^2 w_2$. Finally, if $w_2 = w_1$ then $D_2 \simeq a_o^2 w_2 p_v/2$.

The behavior in a three-dimensional lattice is just the same. If an impurity-vacancy pair is formed in an fcc lattice, the only difference from the above example is that there are four solvent atom sites which the vacancy can move to without dissociating the impurity-vacancy pair, instead of two (see Fig. 3-9). The probability of an impurity-vacancy exchange on the next vacancy jump is thus $w_2/(4w_1 + w_2)$, and the resulting equation for D_2 varies with the ratio w_1/w_2 in a manner similar to that in Eq. (3-16).

For the case in which the vacancy-impurity pair is less tightly bound,

---

[10]To show this, note that the probability that the vacancy and the impurity will exchange in the time element dt is proportional to $w_2 dt$ if $dt \ll 1/w_2$. The probability that the vacancy will make any one of the three possible jumps in dt is proportional to $(2w_1 + w_2)dt$. The probability of the next vacancy exchange being with the impurity is the ratio of these two or $w_2/(w_2 + 2w_1)$.

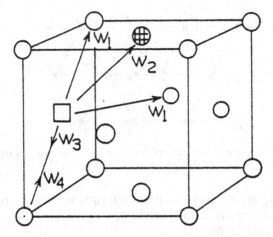

Fig. 3-9—Vacancy-atom jumps used in the 5-frequency model of solute-vacancy interaction in an fcc lattice. w_2 is the solute-vacancy exchange rate. w_1 moves the vacancy to another solute nearest neighbor. w_3 moves the vacancy away from a solute. The reverse of w_3 is w_4. w_o is the jump frequency far from a solute.

the rate of dissociation of the pairs must be included. If we assume that the region affected by an impurity includes only its nearest neighbors, w_3 can be taken as the frequency with which an associated vacancy exchanges with one of the seven solvent atoms which is not a nearest neighbor of the impurity atom, and thus dissociates the vacancy-impurity pair. The probability of an impurity-vacancy pair dissociating on its next jump is found to be $7w_3/(w_2 + 4w_1 + 7w_3)$. If we calculate an approximate value of f as before by assuming $\cos \theta_1 = -w_2/(w_2 + 4w_1 + 7w_3)$, we obtain

$$f = (4w_1 + 7w_3)/(2w_2 + 4w_1 + 7w_3) \qquad (3\text{-}17)$$

A more accurate calculation gives

$$f = (2w_1 + 7w_3)/(2w_1 + 2w_2 + 7w_3)$$

If the calculation is generalized to include the return of dissociated vacancies,

$$f = (2w_1 + 7Fw_3)/(2w_1 + 2w_2 + 7Fw_3) \qquad (3\text{-}18)$$

Substituting this in Eq. (3-14) gives

$$D_2 = a_o^2 p_v w_2 (2w_1 + 7Fw_3)/(2w_1 + 2w_2 + 7Fw_3) \qquad (3\text{-}19)$$

Here two additional frequencies have been added, w_o is the frequency for solvent self-diffusion, and w_4 represents the jump which brings a vacancy to a solute nearest neighbor position; it is the reverse of w_3.

(See Fig. 3-9.) F in Eq. 3-18 is a function of the ratio w_4/w_o with $F = 1$ for $w_4/w_o = 0$ and $F = 2/7$ for $w_4/w_o = \infty$.

The type of complete or partial association of impurities and vacancies discussed above is important in the study of impurity diffusion in ionic materials. For example, if a divalent impurity, such as Mg^{++}, is dissolved in NaCl, charge neutrality requires that the magnesium ion replaces two sodium ions, replacing one with a Mg^{++} ion and leaving the other vacant. Though the vacancy and the Mg^{++} can move separately there will be a strong electrostatic (coulombic) attraction between the two so the fraction of magnesium ions paired with vacancies will be high. Thus it is easy to see why this problem was first treated for the case of impurity diffusion in ionic materials.[11]

This type of approach involves many frequencies, but different types of experiments can be used and by comparing these it is possible to determine the ratios of the five frequencies.[12] The model can also provide useful insights into limiting cases, as can be seen from the following.

- If the rate of dissociation of vacancy-impurity pairs $7w_3$ is much less than $2w_1$, and $w_4 \ll w_o$ so $F = 1$, Eq. (3-19) simplifies to

$$D_2 \simeq a_o^2 w_1 w_2 p_v/(w_1 + w_2) \qquad (3\text{-}20)$$

- If impurity-vacancy exchange is much more rapid, then $w_2 \gg (w_1 + 7w_3F/2)$ and

$$D_2 \simeq a_o^2 p_v(2w_1 + 7w_3F)$$

- If the impurity atom jumps relatively slowly, $w_2 \ll (w_1 + 7w_3F/2)$ and

$$D_2 = a_o^2 p_v w_2$$

- Finally, if the "impurity" is a solvent tracer, then $w_1 = w_2 = w_3 = w_o$ and $p_v = N_v$. Thus

$$D_1 = D_2 = (9/11)a_o^2 N_v w_o \qquad (3\text{-}21)$$

Note that in Eq. (3-22) $f = 9/11 = 0.82$ even though correlation effects were only approximated in the derivation.

Ratio of D's for solvent and solute. If the solute and solvent share the same sites and the same vacancies there clearly are relationships between D_1 and D_2. The first of these is that they cannot differ in magnitude by more than a factor of 10, and will usually be closer than that. Furthermore, it has been observed that if D_1 increases with N_2,

[11]A. Lidiard, *Phil. Mag.*, 46, (1955) 1218.
[12]For example see A. D. LeClaire, *J. Nucl. Matl.*, 69&70 (1978) 70.

then D_2 also increases with N_2. These relationships are often expressed in terms of the equations

$$D_1(N_2) = D_1(0)[1 + b_1 N_2] \quad \text{and} \quad D_2(N_2) = D_2(0)[1 + B_1 N_2] \quad (3\text{-}22)$$

Thus b_1 and B_1 have the same sign, and have a magnitude that can be expressed in terms of the exchange frequencies used above. From the discussion of vacancy-solute binding energies given above, one would expect b_1 and B_1 to be positive for solute to the right of silver in the periodic table, e.g. Sn, and negative for elements to the left, such as Pd. This is indeed the case.[13] In fact the affect of solute can be thought of as adding separate non-interacting regions of a different jump frequency (a higher frequency if b is positive). The observed values of D in the dilute alloy is thus the linear combination of D in the pure metal plus the contribution from the different value in the disturbed regions around the solute atoms.

3.4 INTERSTITIAL DIFFUSION IN SUBSTITUTIONAL ALLOYS

Self-diffusion in pure fcc, hcp, and bcc metals seems to always occur by a vacancy mechanism. Also, the diffusion of substitutional solute in these metals usually occurs by the vacancy mechanism. As was pointed out above, if the solute and solvent share the same vacancies and same lattice sites, then the ratio D_1/D_2 is between 0.1 and 10, since both types of atoms must move for either to allow the diffusion of the other. However, interstitial atoms diffuse much faster than those on substitutional sites and if some of the impurity atoms spend a fraction of their time on interstitial sites, D_2 can be orders of magnitude greater than D_1. Whether a solute goes into solution on interstitial or substitutional sites is usually deduced from whether a smaller diameter solute increases the lattice parameter (interstitial), or decreases it (substitutional). For example an empirical rule based on x-ray data indicates that a solute will dissolve primarily interstitially only if its atomic radius is less than 0.59 that of the solvent. However, diffusion studies discussed below indicate that solute with radii up to 0.85 that of the solvent may spend enough time as interstitials for interstitial motion to dominate the transport of solute.

At thermal equilibrium any impurity will be distributed over both interstitial and substitutional sites. If the interstitial sites are all equivalent then there will be an equilibrium between the normal and interstitial sites, which can be represented by the equation

---

[13]A. D. LeClaire, *J. Nucl. Matl.*, *69&70* (1978) 70–96.

$$s \rightleftarrows i + v \qquad (3\text{-}23)$$

The corresponding atom fractions of these sites are represented by N_s, N_i, and N_v. If these defects are at equilibrium locally then the equilibrium constant can be written

$$N_i N_v / N_s = \exp[-(G_v + G_i)/RT] \qquad (3\text{-}24)$$

where G_i and G_v are the free energies of formation of interstitials and vacancies, respectively. If N_v is determined by an equilibrium with dislocations then $N_v = \exp(-G_v/RT)$ and the fraction of solute atoms present on interstitial sites becomes

$$N_i / N_s = \exp(-G_i/RT) \qquad (3\text{-}25)$$

Reliable calculations indicate that for copper and other close-packed noble metals, G_i is about four times G_v. Thus the concentration of interstitial copper atoms in copper is so small that their contribution to copper self diffusion is negligible. However, there is evidence in a number of systems that solute atoms which lie predominantly on substitutional sites but are at least 15% smaller than the solvent can diffuse many orders of magnitude faster than the solvent atoms. For example the ratio of the atomic radii for gold and lead is 0.83, but the diffusivity of Au in Pb is many orders of magnitude greater than that for Pb in Pb (at 175° C the ratio D_{Au}/D_{Pb} is 10^5). Fig. 3-10 shows the tracer diffusion rates for various elements in lead. It is seen that Cu, Pd, Au, Ni, and Ag all diffuse at least 1000 times faster than Pb in Pb, whereas D for Sn, Tl, Na, Hg and Cd lie within a factor of 10–20 of that for Pb self diffusion. Thus the latter group of elements could diffuse by a vacancy mechanism, while the former, fast diffusing elements could not. Similar results are found for diffusion in tin with a strong similarity in the order of the elements, that is D_2/D_1 decreases in the order Cu, Au, Ag.[14]

The mechanism of this anomalously fast diffusion is not known, though a great deal of sophisticated effort has been devoted to the matter. Using the well studied lead alloys as our example some of the results are presented here to indicate some of the techniques that can be brought to bear to infer diffusion mechanisms.[15] The techniques discussed are centrifugation, isotope effect, and variation of solvent diffusion with solute additions.

Centrifugation. If there are no forces on the atoms in a dilute alloy the equilibrium solute distribution is one with no concentration gra-

---

[14]W. K. Warburton, D. Turnbull, *Diffusion in Solids*, eds. A. S. Nowick, J. J. Burton, Academic Press (1975), p. 171.

[15]This treatment often follows A. D. LeClaire, *J. Nucl. Matl.*, 69&70 (1977) 70–96.

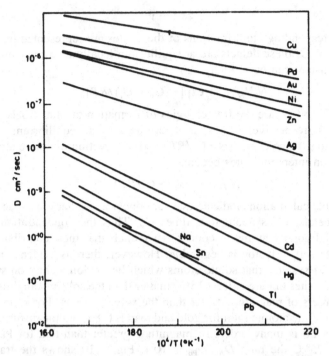

Fig. 3-10—Diffusivities of various elements in lead as a function of temperature. [From W. K. Warburton, D. Turnbull, *Diffusion in Solids*, eds. A. S. Nowick, J. J. Burton, Academic Press (1975), p. 171.]

dient. However, if the sample is placed in a centrifuges it can be given placed under an acceleration in excess of 100,000 g. Under these accelerations small mass differences in the elements of an alloy set up forces that can lead to the development of an appreciable concentration gradient. Whether an object rises or falls in such a column depends on what is being forced to the top of the column if the object sinks. In the case of a Au tracer in Pb what rises depends on the mechanism of diffusion. For example if Pb and Au both diffuse by a vacancy mechanism and occupy lattice sites then since Au is slightly lighter than Pb, the Au would tend to rise since the heavier lead will take the place of any Au initially in the bottom of the column. However, if the Au diffuses by moving over interstitial sites, it will not displace a Pb atom when it sinks to the bottom and there will be a strong tendency for it to sink.

The equation for the actual distribution can be obtained as follows. The centripetal force on a Au atom (subscript 2) can be written as (m_2

$- v'm_1)\omega^2 x$, where x is the distance from the center of rotation of the centrifuge, m_2 and m_1 are the molecular weights of the two atoms, and v' is the ratio of the partial molar volumes of the solute (Au) to that of the solvent (Pb). When this force is inserted in the flux equations of Chap. 1, the equilibrium distribution becomes

$$dln(N_2)/dx^2 = (m_2 - v'm_1)\omega^2/2RT \qquad (3\text{-}26)$$

If Au occupies a normal lattice site then $v' \simeq 1$, so if Au diffuses by a vacancy mechanism the force on the Au atoms will be small and tend to make them move toward the top during diffusion. On the other hand if Au occupied an interstitial position, its partial molar volume would be small, v' would be closer to zero, and in the centrifuge the lead would sink to the bottom of the sample since in doing so it would displace little lead toward the top. The experimental results shown in Fig. 3-11 indicate that the gold clearly sinks toward the bottom, and is consistent with the mobile atoms being interstitials with a finite molar volume, that is v' is somewhat greater than zero. (It has been found that Au sinks in Na and K, while Au in In floats rapidly to the top

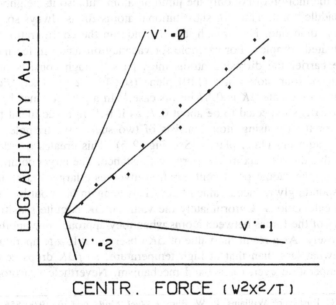

CENTR. FORCE ($\omega^2 x^2 / T$)

Fig. 3-11—Steady-state distribution of gold diffusing in solid lead in a centrifuge at 590° C. Definition of v' given in text. [From S. J. C. Rushbrook Williams, L. W. Barr, *J. Nucl. Matl.*, *69&70* (1978) 556-8.]

and the authors suggest that the gold atom diffuses interstitially in Na and K but in indium diffuses as a gold-vacancy pair.[16])

Isotope Effect. The vibration frequency of a harmonic oscillator is inversely proportional to the square root of its mass. If two different isotopes, a and b, of the same element diffuse through the lattice they will diffuse by the same process. However, the mass of the vibrating complex will differ by the difference of the mass of the two isotopes m_α and m_β and the larger vibration frequency of the lighter atom will lead to its diffusing somewhat faster than that of the heavier atom. The deviation of the ratio of the diffusion coefficients of these two isotopes D_α and D_β from the ratio of the square roots of their masses gives information on the mechanism of diffusion through the equation[17]

$$[(D_\alpha/D_\beta) - 1]/[(m_\beta/m_\alpha)^{0.5} - 1] = f\Delta K \qquad (3-27)$$

Here f is the correlation coefficient. It enters because the increase in the jump frequency due to the lower mass leads to a corresponding increase in D which is less than that of the jump frequency by the correlation factor. The second factor in Eq. 3-27, ΔK is the fraction of the activation energy associated with the motion of the diffusing atom. The vibration mode in a lattice that leads to an atomic jump will involve the motion of not only the jumping atom but also its neighbors at the saddle point. Thus for substitutional atoms ΔK is always somewhat less than one. How much less depends on the configuration of the activated complex. For example for vacancy diffusion in fcc metals, the barrier the diffusing atoms must pass through consists of a rectangle of four atoms in a (110) plane (see Fig. 2-4). Theoretical calculations indicate ΔK is 0.87 in this case.[18] In a pure fcc metal $f = 0.78$ so $f\Delta K$ is expected to be about 0.7, as it is.[19] In a bcc metal the barrier for the diffusing atom consists of two successive triangles of atoms in adjacent (111) planes (See Fig. 2-5). This creates a saddle point with a double maxima in energy vs. distance. The larger complex making up the saddle point configuration involves a larger mass, and a correspondingly reduced value of ΔK. This is again born out by theoretical calculations. Unfortunately the value of ΔK is quite sensitive to values of the forces between atoms when they approach one another quite closely. As a result the value of ΔK observed at low temperature is somewhat less than that at high temperature, so $f\Delta K$ drops some with temperature even for a fixed mechanism. Nevertheless, isotope

[16]S. J. C. Rushbrook Williams, L. W. Barr, *J. Nucl. Matl.*, 69&70 (1978) 556–8.
[17]C. P. Flynn, *Point Defects & Diffusion*, Clarendon-Oxford Press, (1972) 341–9.
[18]C. P. Flynn, personnel communication.
[19]N. L. Peterson, *Diffusion in Solids*, eds. A. Nowick, J. Burton, Academic Press, (1975), pp. 115–68.

effect experiments provide the most direct means for inferring values of f. This suggests the mechanism and indicates whether it changes with temperature in a given system.

If the fast diffusers in lead, such as Cu, Ag or Au, diffused as a simple interstitial atom (like carbon in fcc iron) then $f = 1$ and $f\Delta K \simeq 1$. However, the value of $f\Delta K$ for Ag diffusing in lead is 0.25 suggesting that the mass of the activated complex is much greater that of the Ag atom alone.

Concentration Dependence. The remaining type of evidence used to infer a mechanism is the rate at which D_1 changes as the concentration of solute increases. If $D_2 > D_1$, it is found that $dD_1/dN_2 > 0$. If diffusion occurs by a vacancy mechanism then the coefficient $b_1 = dD_1/dN_2$ can be calculated in terms of the multiple jump frequency model of the preceding section. Qualitatively, if vacancy-solute pairs form and exchange more easily than vacancy-solvent pairs, the addition of solute will increase D_2. However, if diffusion occurs by an interstitial mechanism the diffusion of solute and solvent will be independent of one another, and the ratio D_1/D_2 will not be related to the observed b_1.

This type of argument has been applied to the binary alloys shown in Fig. 3-10. It is found for Hg and Cd in lead that D_1/D_2 and b_1 are directly proportional to another. It is argued that in this case a vacancy-solute interstitial pair forms, and the number of additional effective solvent jumps due to the presence of solute was the same as the effective number of solute jumps.

For Au and Cu in lead the change in dD_1/dN_2 is orders of magnitude less than would be required if the solute diffusion was to occur by a vacancy, or a vacancy-interstitial pair mechanism. For these elements it is believed that an interstitial mechanism operates, but due to the small isotope effect for these solutes it is believed that the interstitial solute and a solvent atom share the same site (as in Fig. 2-6.), and both must move in an activated jump. If the concentration of the solute in interstitial sites is c_i and that in substitutional sites is c_s, and the corresponding diffusion coefficients in these two types of sites are D_i and D_s, then the equation of D_2 is a weighted average of the jumps that comes from each of the two types of sites, or

$$D_2 = [c_i/(c_i + c_s)]D_i + [c_s/(c_i + c_s)]D_s \qquad (3\text{-}28)$$

Estimates are that D_i is over 10,000 times D_s, so c_i can be much less than c_s and still have the dominant transport mechanism be due to interstitial diffusion.

The principal conclusion to be drawn from this is that solute atoms diffuse by a variety of mechanisms in lead and tin. Similar results are found in some bcc metals, and have been interpreted in this manner, though the situation there is more controversial.

3.5 DIFFUSION WITH TRAPS (HYDROGEN IN IRON)

Hydrogen being the smallest atom might be expected to diffuse more easily through metals than any other element, and it does. Fig. 3-12 shows D for Fe, C and H in α-Fe from 1000° K to below room temperature. Note that at 300° K an iron atom changes position once per 100 yr, and carbon every 10 s, while hydrogen is still jumping at a rate of $10^{12}/s$. Even at low concentrations hydrogen often leads to the embrittlement of metals, for reasons that are poorly understood, but certainly are related to the speed at which it can diffuse to highly stressed regions.

The main purpose of this section is to indicate the large effect that trapping at defects can have on diffusion in solids of solute with a low equilibrium solubility. Hydrogen diffusion is used as an example since it diffuses so easily that even shallow traps will produce a measurable effect on D.[20] Hydrogen in alpha iron is the most commonly studied system showing this type of behavior. The extrapolation of high temperature solubility (above 400° C) to low temperatures indicates that the solubility of hydrogen in ferrite at room temperature would be about one part in 10^8, by weight. However, the observed solubility at room temperature can be much greater than this, the exact value depending on the density of low energy sites introduced into the lattice by dislocations, ferrite-cementite interface, microvoids, inclusions, etc. These low energy sites serve as traps which inhibit the diffusion of hydrogen.

Consider two sheets of iron at room temperature with a hydrogen gradient across them. One sheet contains no defects while the other contains many dislocations. In the perfect crystal the hydrogen moves with an activation energy of 8 kJ/mol. In the imperfect crystal if a hydrogen atom moves to a dislocation the energy of the lattice is reduced by about 0.2 eV/atom, that is there is a binding energy of about 25 kJ/mol which the hydrogen atoms must overcome before it can escape from the dislocation and diffuse away through the lattice.

To obtain an equation relating the effective diffusion coefficient D_e to the diffusion coefficient in the perfect lattice D_L, consider the conservation of matter. Since hydrogen is either in traps or perfect lattice sites, the rate of change of total concentration is given by the equation

---

[20]The literature on hydrogen in metals is immense. J. P. Hirth, *Met. Trans.A*, *11A* (1980) 861–90, summarizes the behavior of H in Fe. H. H. Johnson, *Met. Trans.A*, *19A* (1988) 2371, treats diffusion and trapping.

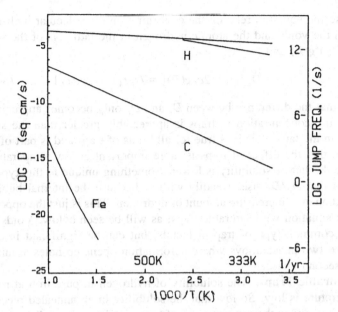

Fig. 3-12—D and jump frequency for hydrogen, carbon and iron atoms in alpha iron between 1000° K and room temperature.

$$\frac{\partial c}{\partial t} = \frac{\partial c_L}{\partial t} + \frac{\partial c_t}{\partial t} = D_L \nabla^2 c_L$$

If we assume equilibrium exists between H atoms on trap sites and lattice sites, i.e. the thermal energy is enough for the atoms to jump out of the traps, then there will be an equilibrium relationship between c_t and c_L of the form $c_t = f(c_L)$, and the two time derivatives can be combined to give[21]

$$\frac{\partial c_L}{\partial t} = \frac{D_L}{1 + f'} \nabla^2 c_L \quad \text{where} \quad f' = dc_t/dc_L \qquad (3\text{-}29)$$

Thus the effective diffusion coefficient can be defined as

$$D_e = D_L/(1 + f') \qquad (3\text{-}30)$$

Bubbles as Traps. As the simplest example of trapping consider the case of hydrogen trapped as gas molecules in internal voids. The solubility of hydrogen usually increases as the square root of the pres-

---

[21]The characterization of traps from data on the variation of D with lattice concentration is developed by, H. H. Johnson, N. Quick, A. J. Kumnick, *Scripta Met.*, **13** (1979) 67–72.

sure so the equation relating the concentration of molecular hydrogen gas in the voids and the atomic hydrogen in the lattice is of the form $c_t = (c_L)^2 g(T)$ and

$$D_e = D_L/[1 + 2c_L g(T)] = D_L/[1 + 2c_t/c_L] \qquad (3\text{-}31)$$

Note that the difference between D_e and D_L only becomes appreciable when the concentration in traps is appreciably greater than the solubility in the lattice. This is true for all kinds of traps and is part of the reason that the effect of traps is more important at low temperatures where the lattice solubility is lower. Something unique to this type of trap is that D_e/D_L rises steadily with c_L because the internal bubbles collect an ever increasing amount of hydrogen. This is just the opposite of the situation with saturable traps, as will be seen below. Voids are not a common type of trap in metals, but can be significant in cold worked two-phase alloys where deformation opens up holes around a hard second phase.

Saturable Traps. The solubility of hydrogen in pure iron at room temperature is low. So low that the solubility in an annealed piece is much lower than the concentration of trapping sites around dislocations in a cold worked piece. In such a case the diffusion of hydrogen at low hydrogen concentrations is determined not by the jump frequency of hydrogen in the perfect lattice, but by the escape frequency of hydrogen atoms from the traps, e.g. low energy sites around dislocations. As a result the effective diffusion coefficient of hydrogen in the hydrogen-free cold worked metal is much lower than D in a well annealed specimen. However, there are only a limited number of low energy sites around a dislocation, and if enough hydrogen is added to fill all the traps, the hydrogen diffusion coefficient in the cold worked crystal increases to that of the annealed crystal. This pronounced effect of trap density and hydrogen content has lead to a wide range of reported values for D_H in iron and steel, and is reflected in Fig. 3-12 by a pair of lines for D_H at temperatures below 100° C.

The equation for saturable traps is more complicated than Eq. (3-31). For simplicity we assume:

- only one type of trap exists and it has a binding energy H_b for hydrogen that is independent of the fraction of traps filled,
- trap sites can only hold one hydrogen atom, i.e. they are said to be saturable,
- equilibrium exists between H atoms on trap sites and lattice sites, i.e. the thermal energy is enough for the atoms to jump out of the traps,
- the atom fraction of sites that are trap sites is much less than unity.

These assumptions are needed primarily to give a relation between c_L and c_t. In a system in which the hydrogen atoms are at equilibrium between the two types of sites a fraction θ_t of the trap are occupied, θ_L of the lattice sites occupied, and the enthalpy difference between the two types of sites is H_b. Consideration of only the ideal entropy of mixing on the two types of sites gives

$$\theta_t/(1 - \theta_t) = \theta_L \exp(-H_b/RT) = \theta_L K \qquad (3\text{-}32)$$

where the assumption $\theta_L \ll 1$ has been used, and the entropy change S_b is neglected. Note that when θ_t is small the fraction of trap sites filled is directly proportional to the lattice concentration, $\theta_t \simeq \theta_L \exp(-H_b/RT)$. The fact that each trap site can hold only one hydrogen atom leads to the $(1 - \theta_t)$ term in Eq. (3-32) which limits the fraction of traps that can be filled. That is, there is a steady increase in the entropy of mixing for adding hydrogen to the last half of the vacant traps.

The atom fraction of lattice and trap sites is related to the number of such sites per unit volume, N_L and N_t, by the equations $c_L = N_L\theta_L$ and $c_t = N_t\theta_t$. Substitution in Eq. (3-32) gives

$$c_t[1 + Kc_L/N_L] = N_tKc_L/N_L \qquad (3\text{-}33)$$

substitution of dc_t/dc_L in Eq. (3-29) gives[22]

$$D_e = D_L\left[1 + \frac{N_tN_LK}{(N_L + Kc_L)^2}\right]^{-1} = \left[\frac{D_Lc_L}{c_L + c_t(1 - \theta_t)}\right] \qquad (3\text{-}34)$$

Note that when $\theta_t \simeq 1$,

$$D \simeq D_L, \quad \text{and} \quad R d\ln D/d(1/T) = Q = H_m$$

while when $\theta_t \ll 1$,

$$D_e = D_L(c_L/c_t), \quad \text{and} \quad R d\ln D/d(1/T) = Q = H_m + H_b,$$

Again a strong variation of D_e requires that $N_t \gg N_L$. Also, the most rapid change of D_e with c_L is in the range of $0.1 < \theta_t < 0.9$.

As an example of the phenomena that can be explained with this model, consider several observations made by Darken and Smith.[23] They studied the uptake of hydrogen by cylinders of annealed, and cold worked, steel with 0.2% C. In the annealed steel the carbide/ferrite interface provides the dominant trapping site for hydrogen. In iron at room temperature the lattice solubility of hydrogen is so low that $N_t \gg N_L$ even in annealed steel. Their observations were:

[22]R. A. Oriani, *Acta Met.*, 18 (1970) 147.
[23]L. S. Darken, R. P. Smith, *Corrosion*, 5 (1949) 1–16.

- The amount of hydrogen pickup by the steel cylinders initially increased as $\sqrt{t}$, and the time to saturate the cylinder increased as the diameter of the cylinder squared. Both of these observations indicate that the hydrogen pickup of the cylinders is diffusion controlled, that is Fick's Laws are obeyed.
- The solubility of hydrogen in steel for a given potential maintained on the surface could be obtained by immersing it in dilute acid so that hydrogen diffused in from both sides and saturated the sheet. The concentration so measured is designated C_2. In a second experiment acid was placed on one side of the sheet and a vacuum on the other side. After steady-state was attained, the average concentration of hydrogen from this one sided charging was measured and designated C_1. If D was a constant independent of concentration, the ratio C_1/C_2 would equal 0.5. For the annealed steel it was found that $C_1/C_2 = 0.8$. Drawing curves of concentration vs. distance through the sheet should indicate to the student that such a ratio will only be found if D decreases appreciably in going from the high hydrogen side of the sheet to the other. Thus a plot of $c(x)$ is curved and concave downward.
- The time to saturate a cylinder decreased as the hydrogen potential of the charging acid increased. That is, if the concentration of hydrogen maintained in the metal at the surface was low, the time to saturation was longer than if the surface concentration was high. (This is contrary to what is predicted by the solutions for constant D given in Chap. 1.) Also, using a high surface concentration, they found the time to saturate a cylinder was appreciably shorter than the time required for the hydrogen to diffuse out into a vacuum. Both of these observations are consistent with the increase of D with hydrogen concentration.
- Cold work increased the solubility of hydrogen in the steel by an order of magnitude for a given hydrogen potential. This is due to the increase in trap density with cold work. Much of this increase is due to trapping at dislocations, though some may be due to voids which open up in the carbide or carbide/ferrite interface of the pearlite upon deformation.

'Irreversible' Traps. In the preceding paragraphs it was assumed that there was only one kind of trap, and that H_b/RT was small enough that there was equilibrium between hydrogen in lattice sites and trap sites. We now consider cases in which each of these assumptions is no longer true. Irreversible trapping is a phenomenon which can easily be observed in metals because the activation energy for lattice diffusion H_m is small relative to the binding energy, H_b, for hydrogen at the strong (irreversible) traps. Thus hydrogen may diffuse easily into iron at room temperature but once trapped be bound so tightly at traps that it cannot escape, that is, $H_b \gg RT$. If the sample is subsequently heated in a vacuum, the rate of escape from these deep traps determines the rate of escape from the sample. This escape is thermally activated and taken to be proportional to $\exp(-H_a/RT)$, where the activation energy for release is approximated by $H_a = H_b + H_m$, as indicated in Fig. 3-13. If the rate of release is taken to be proportional to the concentration in the traps c_t, the release rate is described by the equation

$$dc_t/dt = -Ac_t \exp(-H_a/RT) \tag{3-35}$$

If the charged sample is continuously heated from a low temperature at a rate ϕ, the rate of evolution is initially zero since $(H_a \gg RT)$. Evolution will begin when RT approaches H_a and increase rapidly until most of the traps are emptied, that is c_t goes to zero. The temperature of the peak in the evolution rate, T_p, is obtained by setting the differential of Eq. (3-35) with respect to t equal to zero, and is given by

$$\phi H_a/R(T_p)^2 = A \exp(-H_a/RT_p) \tag{3-36}$$

If there are two types of traps with different binding energies, the hydrogen traps will release hydrogen independently of one another and peaks will appear at two different temperatures. Using this technique

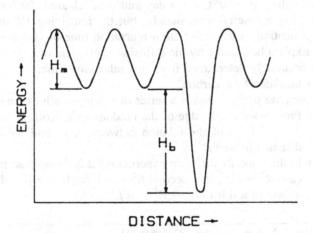

Fig. 3-13—Schematic diagram of the energy level of hydrogen in iron showing the energy for lattice diffusion H_m, and the binding energy H_b.

Lee et al.[24] have studied traps for hydrogen in α-Fe. Values for H_a are 18.5 kJ/mol at Fe/Fe$_3$C interface, 26.9 at dislocations, and 86.9 at the Fe/TiC interface.

Further examples of diffusion with internal traps:

- Internal oxidation of metals is another case of irreversible trapping of a more rapidly diffusing solute, this time oxygen combines with a reactive solute, such as aluminum in silver.[25]
- In amorphous materials like glass, hydrogen solubility and diffusivity indicate the energies of the sites the hydrogen can occupy are best expressed as a continuous distribution of energies (trap depths) around a mean.[26]

PROBLEMS

3-1. An internal friction peak caused by interstitial diffusion in a bcc metal has its peak at 50° C at a frequency of 0.70 Hz. When the frequency is changed to 2.81 Hz the peak is shifted to 64° C.
  (a) Calculate Q for the process.
  (b) If the lattice parameter is 3.2×10^{-8} cm, calculate D at 100° C and D_o.

3-2. Draw a fcc unit cell, and explain why individual interstitial atoms do not give rise to an internal friction peak. Could interstitial pairs give rise to internal friction?

3-3. A small stress (under 1% of the yield stress) is applied to a dilute Fe-C alloy at −50° C for a day and then released. 99% of the strain is recovered immediately, but the remaining 1% decays exponentially with time with a relaxation time of 100 min.

  Explain how and why the diffusion coefficient of C in Fe can be accurately determined from this relaxation time.

  Calculate D for carbon.

3-4. A jumping particle makes a series of n jumps each of length L.
  (a) From your knowledge of the random walk problem write a general form of the relation between n, L, and the mean distance moved R^2.
  (b) In three totally different experiments it is found that: in one case $R^2 = nL^2$, in a second $R^2 = 0$ though $n \gg 0$ and $L > 0$, and in a third $nL^2 < R^2 < n^2L^2$.

[24]H. G. Lee, J. Y. Lee, Acta Met., 32 (1984) 131–6.
[25]N. Birks, G. H. Meier, Introduction to High Temperature Oxidation of Metals, E. Arnold, (1983), p. 95.
[26]R. Kirchheim, F. Sommer, G. Schluckebier, Acta Met., 30 (1982) 1059–68.

Explain the different relationships that must exist between the successive jump directions for each of the three cases.

3-5. Would you expect any correlation between the successive jump vectors of a tracer in a pure metal if diffusion occurs by
   (a) A vacancy mechanism
   (b) An interstitial mechanism
   (c) An interstitialcy mechanism

3-6. Calculate $\overline{\cos(\theta_1)}$ and f for a tracer diffusing in a two-dimensional square lattice by a vacancy mechanism. (Consider only vacancy exchanges with nearest neighbors, and trajectories which move the solute again in three or fewer vacancy jumps.)

3-7. The equilibration time for the diffusion of hydrogen into a piece of iron initially free of H was measured at 50° C using two different hydrogen pressures, one quite low, and one quite high. The time to reach saturation was much longer for the low pressure than for the high. Similar tests on thinner samples gave the same relative rates, thus a slow solution step at the surface was ruled out as rate determining.
   (a) Explain why saturation took longer when the ambient pressure of hydrogen was low.
   (b) This difference in apparent D between high and low hydrogen pressure disappears if the same experiment is performed at 200° C. Why?

3-8. Steady-state diffusion of hydrogen is established through a sheet of metal with a high hydrogen concentration maintained on one side and a concentration of zero on the other side. Plot $C(x)$ through the sheet for the following three cases:
   (a) D constant independent of C.
   (b) D increasing by a factor of 10 from low to high C.
   (c) D decreasing by a factor of 10 from low to high C.
   (d) For each of the variations of $D(C)$ considered above, compare the total content of the sample to that of one saturated at the high value of C.

3-9. A cold worked sample of steel is charged with hydrogen at room temperature, heated at a rate of 0.1° C/s, and the rate of hydrogen evolution measured.
   (a) Calculate the temperatures of the maximum hydrogen release rates for a sample with two types of traps having binding energies of 18.5 kJ/mole and 26.9 kJ/mol. Take $A = 0.01/s$ in Eq. 3-36.
   (b) What would the temperatures be for the two types of traps if the heating rate was decreased by a factor of ten?

3-10. The evolution of hydrogen from a charged sample may be limited by either lattice diffusion or escape from traps, depending on the relative values of H_m the activation energy for diffusion, and H_b the binding energy at the traps.

(a) Two identically charged samples, differing only by a factor of two in thickness, are quickly heated to a given temperature and then held there. How will the time to evolve 90% of the hydrogen $t(.9)$ differ between the two samples if the evolution is controlled by lattice diffusion?

(b) If escape from strong traps rather than diffusion limits hydrogen evolution in the samples described in (a), how will $t(.9)$ differ for the two samples?

(c) If the two charge sample are heated at a slow uniform rate, draw a plot of hydrogen evolution rate for the case in which lattice diffusion limits evolution, and another for the case in which evolution from traps limits the release rate.

Answers to Selected Problems

3-1. (a) 21.3 kcal/mol (89.2 kJ/mol).   (b) $D_o = 0.033$ cm^2/s.

3-3. 1.3×10^{-20} cm^2/s

3-5. (a) yes,   (b) no,   (c) yes, for every other jump.

3-6. $f = 0.542$

3-7. (a) The rate of penetration is proportional to $D_L c_L / c_t$. Since c_L increases as $P^{1/2}$ while D_L and c_x are independent of P, the rate is lowest at low pressure.

(b) c_L increases substantially with temperature while $c_L = c_L + c_t$ will decrease due to the drop in c_t with rising T.

3-8. (d) $(\bar{c}/c_{max}) = 1/2$ for (a), is about 2/3 for (b) and about 1/3 for (c).

3-9. Using Eq. (3-36), (a) 712° K. 963° K. (b) 480° K, 660° K

3-10. (a) $t(.9)$ will differ by a factor of 4 for the two.

(b) $t(.9)$ will be much greater than for diffusion control, and will have the same value for the both thicknesses.

(c) For diffusion control, the maximum rate of evolution will be arise at a higher temperature (T_{max}) for the thicker sample. For trap release control, T_{max} will be the same for both samples, but the rate will be higher for the thicker sample.

4

DIFFUSION IN A CONCENTRATION GRADIENT

In the preceding chapter our discussion of diffusion in substitutional alloys was limited to self-diffusion experiments. In such experiments the specimen is, or is assumed to be, chemically homogeneous. Such studies showed that the self-diffusion coefficients are, in general, different for the two elements in a substitutional alloy. Yet, if two semi-infinite bars of differing proportions of components 1 and 2 are joined and diffused, the Boltzmann-Matano solution gives only one diffusion coefficient $D(c)$ which completely describes the resulting homogenization. Thus the problem is to relate this single diffusion coefficient to the self-diffusion coefficients at the same composition. To do this two new effects must be understood. The first of these concerns the kind of matter flow which is to be classified as diffusion. In a binary diffusion couple with a large concentration gradient we shall see that diffusion gives rise to the movement of one part of the diffusion couple relative to another. The coordinate system used in the Boltzmann-Matano solution is fixed relative to the end of the specimen, and the chemical diffusion coefficient is given by the equation[1]

$$\bar{D} = -J/(\partial c/\partial x) \qquad (4\text{-}1)$$

Thus any movement of lattice planes relative to the ends of the dif-

---

[1]In this chapter it is necessary to work with several different diffusion coefficients, all of which apply to the same system but which are different. We shall define each of them by an equation. The D obtained from the Boltzmann-Matano solution is called the chemical diffusion coefficient and will be designated $\bar{D}$.

fusion couple is recorded as a flux and affects $\bar{D}$ even though such translation does not correspond to any jumping of atoms from one site to another.

The second effect deals with the relation between the diffusion coefficient measured in a tracer experiment and the intrinsic diffusion coefficient of the separate elements in a binary diffusion couple. This requires a more detailed consideration of the chemical forces giving rise to diffusion in solids. The development will be based primarily on thermodynamic or phenomenological reasoning as opposed to the atomistic models of Chaps. 2 and 3.[2]

4.1 THE KIRKENDALL EFFECT

Intuitively it would seem that $\bar{D}$ is some kind of mean value of the diffusion coefficients of components 1 and 2. But, if components 1 and 2 diffuse at different rates in a binary alloy, it is necessary to obtain some parameter other that $\bar{D}$ to indicate the magnitude of this difference. The first experiment which allowed the determination of this difference for alloys was discovered by Kirkendall.[3] In the experiment used, a rectangular bar of 70-30 brass was wound with fine molybdenum wire (molybdenum is insoluble in copper and brass) and then plated with about 0.1 in. of copper. This couple was then given a series of successive anneals. After each anneal, a piece was cut from the bar, polished, and the distance between the Mo wires (d) was measured (see Fig. 4-1). It was found that d decreased an amount proportional to the square-root of time. In the Cu-Zn system there is a small volume change on adding copper to brass, but even after this effect was subtracted out, a definite marker shift remained. This shift required that the flux of zinc atoms outward past the markers be appreciably greater than the flux of copper atoms inward across the same plane. Kirkendall had attempted to show this effect in two earlier papers, and in this case the results were sufficiently unequivocal to move his peers. In 1947 this was a new concept,[4] and its generality was not apparent. However, later work on a variety of markers in many different alloy systems confirmed these results, and the effect has proved to be quite general.[5]

---

[2]For a summary of diffusion data in alloys see *Smithells Metals Ref. Book*, 6th Ed., Butterworth (1983), Tables 13.3 and 13.4.

[3]A. Smigelskas and E. Kirkendall, *Trans. AIME*, *171* (1947) 130.

[4]The reader who is interested in the change which this made in concepts of alloy diffusion will find the discussion of Smigelskas and Kirkendall's paper interesting reading.

[5]The work of L. C. Correa da Silva and R. Mehl, *Trans. AIME*, *191* (1951) 155, is one of the careful, early confirmations. For a list of such references, see D. Lazarus, in Seitz and Turnbull (eds.), *Solid State Physics*, vol. 10, p. 71, Academic Press, New York, (1960).

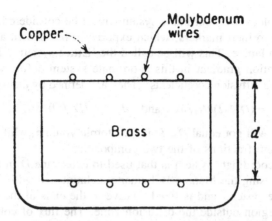

Fig. 4-1 — Schematic diagram showing a cross section of the diffusion couple used by Smigelskas and Kirkendall.

4.2 DARKEN'S ANALYSIS

In 1948 Darken published an analysis of diffusion in alloys which was inspired by the experiments of Smigelskas and Kirkendall.[6] In it he established answers to the question of how the tracer diffusion coefficients are related to $\bar{D}$ and the nonideality of the alloy. Darken's original paper is an excellent example of a phenomenological analysis. A basic characteristic of this approach is that no atomistic model is assumed so that the results are quite general. We will forego some of this generality by assuming a vacancy mechanism and thereby work with a more specific model.

It is useful to begin by considering what type of atomic motion is to be considered as diffusion. The marker movement experiments show that the region around the markers translates relative to the ends of the sample where no diffusion occurs. The uniform translation or flow of an entire region across a reference plane gives a flux through the plane, but this is not what would normally be termed a diffusion flux across the plane. The concepts of flow may be clearer if the problem of the diffusion of ink in a moving stream of water is considered. If water containing a concentration of ink c flows past a point on the bank at a velocity v, the flux of ink past that point will be vc plus any flux due to a concentration gradient in the water. To separate these two contributions to the flux, it is necessary to determine v. To do this chips of wood could be placed on the water and their velocity taken to be that of the water.

[6]L. Darken, *Trans. AIME*, *175* (1948) 184.

In this problem two coordinate systems must be considered. One is fixed relative to inert markers, which experience has shown are fixed relative to the lattice. This will be called the 'lattice system'. The flux and concentration gradient in this coordinate system define what are called 'intrinsic diffusion coefficients'. They are defined by the equations

$$J_1 = -(D_1/\Omega)\partial N_1/\partial x \quad \text{and} \quad J_2 = -(D_2/\Omega)\partial N_2/\partial x \quad (4\text{-}2)$$

D_1 in general will not equal D_2. Ω is the atomic volume, while N_1 and N_2 are the atom fractions of the two components.

The other coordinate system is that used to determine $\bar{D}$ in a binary alloy couple using the Matano-Boltzmann technique. It will be called the 'reference system' and is fixed relative to the ends of the sample, that is in a region outside the diffusion zone. The flux of component 1 in this system, which we will designate Jr_1, plus that for component 2, Jr_2, equals zero, or

$$Jr_1 + Jr_2 = 0 \quad (4\text{-}3)$$

The diffusion coefficient in such a system is given by Eq. 4-1. It is further assumed that the volume of the couple does not change during diffusion. This is equivalent to assuming that the molar volume of the two species is independent of composition.

We now wish to obtain equations which interrelate the velocity of these two coordinate systems relative to each other (the marker velocity) and the various diffusion coefficients $\bar{D}$, D_1 and D_2.

In general $D_1 \neq D_2$. Since $-\partial N_1/\partial x = \partial N_2/\partial x$ it follows that in the lattice system more material will diffuse out of one side of the couple than diffuses in. Though the equations obtained are independent of the mechanism operating, a vacancy mechanism of diffusion will be assumed here to aid in visualizing the situation. In regions without sources or sinks (the great majority of the lattice), the number of lattice sites is fixed, and the sum of the fluxes of atoms and vacancies in the lattice coordinate system is zero,

$$J_v + J_1 + J_2 = 0 \quad (4\text{-}4)$$

Clearly $J_v = -(J_1 + J_2)$ will not be zero. In metals it is a good approximation to assume that the vacancy concentration keeps its equilibrium value at each point. If this is true, all of the vacancies that flow through the couple must be created on one side of the couple and annihilated on the other. This will move the markers relative to the ends of the sample. The marker velocity v will equal the net flow of atoms, which is the vacancy flux times the atomic volume

$$v = J_v \Omega = -(J_1 + J_2)\Omega = (D_1 - D_2)\partial N_1/\partial x \quad (4\text{-}5)$$

To relate $\bar{D}$ to D_1 and D_2 note that the flux of component 1 in the reference system is given by the following equation ($J\Omega$ here has dimensions of length/time)

$$Jr_1\Omega = \Omega J_1 + vN_1 = \Omega[J_1 - N_1(J_1 + J_2)]$$

$$= \Omega(N_2 J_1 - N_1 J_2)$$

$$= -(N_2 D_1 + N_1 D_2)\, \partial N_1/\partial x \qquad (4\text{-}6)$$

But comparing this equation to Eq. 4-1 shows that the desired relationship for $\bar{D}$ is

$$\bar{D} = N_2 D_1 + N_1 D_2 \qquad (4\text{-}7)$$

If the equation for Jr_2 is used in Eq. 4-6, the same Eq. 4-7 is obtained. This is thus a 'proof' that there is only one diffusion coefficient in the reference coordinate system, as was asserted at the start of this chapter.

Equations (4-5) and (4-7) then completely describe the results for the case of isothermal diffusion in an infinite couple. The treatment simply states that if a marker movement is observed, the magnitudes of D_1 and D_2 are different and can be determined from measurements of v and $\bar{D}$.

As an example of the use of these equations for determining D_1 and D_2, we shall analyze the results of Smigelskas and Kirkendall. Empirically it was found that the distance between the molybdenum wires placed at the copper-brass interface decreased as $\sqrt{t}$. If this is true, then the distance the wires have moved from their initial position (x_m) is given by the equation

$$x_m = \alpha\sqrt{t} \qquad (4\text{-}8)$$

or[7]

$$v = dx_m/dt = x_m/2t \qquad (4\text{-}9)$$

In an 'infinite' diffusion couple any given composition shifts as $\sqrt{t}$, so the markers stay at the same composition.[8] The values of D_{Cu} and D_{Zn} can thus be determined for this composition. For diffusion couples of Cu against 70/30 brass Smigelskas and Kirkendall found the marker composition to be 22.5% Zn, and after 56 days at 785° C the markers shifted 0.125 mm toward the brass. Using this and their other data,

---

[7]Equation (4-9) is valid only for markers placed at the initial interface, while Eq. (4-5) is valid for a marker no matter where it was initially placed.

[8]This follows from the substitution made in the Boltzmann analysis for an infinite couple (Sec. 1-6) in which it was shown that $c = f(\lambda) = f(x/2\sqrt{Dt})$. Thus any c will correspond to a constant value of λ, but a point of constant λ is given by the equation $x = (\text{const})\sqrt{Dt}$.

Darken calculated that at 22.5% Zn

$$D_{Cu} = 2.2 \times 10^{-9} \text{ cm}^2/\text{s}$$

$$D_{Zn} = 5.1 \times 10^{-9} \text{ cm}^2/\text{s}$$

$$D_{Zn}/D_{Cu} = 2.3$$

As N_{Zn} approaches zero, Eq. (4-7) indicates that $\tilde{D}$ approaches D_{Zn}. From earlier work by Rhines and Mehl, $\tilde{D} = D_{Zn} = 3 \times 10^{-10} \text{ cm}^2/$ s (at 0% Zn). This indicates that D_{Zn} increases seventeenfold in going from 0 to 22.5% Zn, and is consistent with the magnitude of increase discussed in Chap. 3. However, as will be seen below this value of D_{Zn} cannot be equated directly to the diffusion coefficient obtained from a tracer experiment in the absence of a concentration gradient. Figure 4-2 shows a concentration-distance curve for a diffusion couple consisting of 90 Cu/10 Zn against 70 Cu/30 Zn. Note the following:

• there is substantial asymmetry in $c(x)$ with much deeper penetration on the high Zn side
• the composition at the marker plane is again at 22.5% Zn
• the marker motion is given by the difference between the marker plane after diffusion and the Matano Interface (Sect. 1-6).

Our calculation of D_{Zn} and D_{Cu} in no way acts as a check on the validity of Darken's analysis. To do this one must compare the results with experiments, and consider the assumptions leading to Eqs. (4-7) and (4-5). The assumption in this derivation that the molar volume of

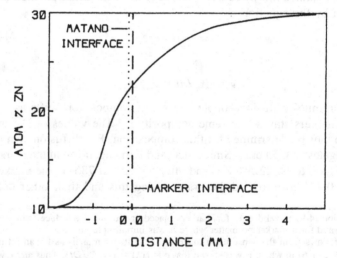

Fig. 4-2—$C(x)$ for 10Zn vs. 30Zn brass couple diffused 144 h at 887° C. [After G. T. Horne, R. F. Mehl, *Trans. AIME*, *203* (1955) 88–99.]

the alloy does not change with composition is a minor one, and can be corrected for in a straightforward manner.[9] The assumption that the flux relative to the markers is given by $-D_1(\partial C_1/\partial x)$ is more fundamental and more difficult to evaluate. To discuss it properly, and to compare D_1 and D_2 with the experimental values of the tracer diffusion coefficients in the same alloy requires a more general description of the flux equation and the forces giving rise to diffusion.

4.3 PHENOMENOLOGICAL EQUATIONS

In chap. 1 it was stated that in a binary phase, if the absence of a concentration gradient is an adequate condition for equilibrium, one is safe in using Fick's first law as a flux equation since the flux will go to zero as the system approaches equilibrium. This is applicable to many systems, and since the concentration is easily measured, it is commonly used. However, $\partial c/\partial x = 0$ is a very restricted condition for equilibrium. To gain insight into just what the limitations of this condition are, it is necessary to use a more general condition for equilibrium.

For a given n-component system at equilibrium, the system can be uniquely determined by specifying T, P, μ_1, μ_2, ..., μ_{n-1}, and ϕ, where μ_i is the chemical potential[10] and ϕ is any relevant scalar potential, e.g., electric potential. If now the system is displaced slightly from equilibrium, it seems most likely, and is certainly simplest, to assume that the rate of return to equilibrium is proportional to the deviation from equilibrium. And, until it is proved to be unnecessary, the flux of, say, component 1 is assumed to be proportional to the gradient of each of the potentials listed above. Thus the most general equation for J_1 is

$$J_1 = -L_{11}(d\mu_1/dx) - L_{12}(d\mu_2/dx) - \ldots - L_{1n}(d\mu_{1n}/dx)$$

$$- L_{1q}(dT/dx) - L_{1P}(dP/dx) - L_{1e}(d\phi/dx) \qquad (4\text{-}10)$$

Similar equations for the flux of component 2, the flux of heat, etc., can be given. These are called phenomenological equations since they stem from no model, but from the observed conditions of equilibrium. A general discussion of the derivation of these equations is given by

---

[9]R. W. Balluffi, *Acta Met.*, 8 (1960) 871.
[10]The chemical potential μ_i is defined by the equation: $\mu_i = (\partial G/\partial n_i)_{P,T,nj}$ $i \neq j$. Where G is the Gibbs free energy of the subsystem or phase. The minimization of G for the system, at constant P and T, is equivalent to the requirement that there be no gradients in μ_i in the system.

Flynn.[11] Experimentally it is found that L_{1q}, L_{1e}, and L_{1P} are not zero, but for now we are interested only in the fact that in an isothermal, isobaric, isopotential system, J_1 is not only proportional to $d\mu_1/dx$, but is also proportional to $d\mu_2/dx$, $d\mu_3/dx$, etc., as well. A complete, concise discussion of the application of the phenomenological equations to alloy diffusion is given by Howard and Lidiard.[12] We shall discuss here only the assumptions required to obtain the flux equation assumed by Darken.

Consider a one-dimensional diffusion problem in an isothermal two-component system. For generality, separate equations are written for the flux of vacancies J_v as well as J_1 and J_2.[13]

$$J_1 = -L_{11}(d\mu_1/dx) - L_{12}(d\mu_2/dx) - L_{1v}(d\mu_v/dx) \qquad (4\text{-}11)$$

$$J_2 = -L_{21}(d\mu_1/dx) - L_{22}(d\mu_2/dx) - L_{2v}(d\mu_v/dx) \qquad (4\text{-}12)$$

$$J_v = -L_{v1}(d\mu_1/dx) - L_{v2}(d\mu_2/dx) - L_{vv}(d\mu_v/dx) \qquad (4\text{-}13)$$

In any region where lattice sites are neither created nor destroyed, the three fluxes are related by Eq. (4-4). If this is to be true for any value of each of the gradients, substitution of Eqs. (4-11) to (4-13) into (4-4) shows that we must have

$$L_{11} + L_{21} + L_{v1} = 0,$$

$$L_{12} + L_{22} + L_{v2} = 0,$$

$$L_{1v} + L_{2v} + L_{vv} = 0$$

For the equations as written here, there is a set of reciprocity relations due to Onsager which state that $L_{ij} = L_{ji}$. Thus

$$L_{12} = L_{21}, \quad L_{v1} = L_{1v}, \quad L_{v2} = L_{2v}.$$

Using all the equations between the L_{ij} gives

$$J_1 = -L_{11}\, d(\mu_1 - \mu_v)/dx - L_{12}\, d(\mu_2 - \mu_v)/dx \qquad (4\text{-}14)$$

$$J_2 = -L_{21}\, d(\mu_1 - \mu_v)/dx - L_{22}\, d(\mu_2 - \mu_v)/dx \qquad (4\text{-}15)$$

These are now the simplest equations which apply in general for a binary alloy where a vacancy mechanism is allowed. If a vacancy mechanism does not operate, the vacancy concentration will be at equilibrium at all points, and $d\mu_v/dx$ will equal zero (as will μ_v).

---

[11]C. P. Flynn, *Point Defects and Diffusion*, Clarendon Press, Oxford, 1972, Chap. 5.

[12]R. E. Howard, A. B. Lidiard, *Rep. Prog. Phys.*, 27 (1964) 162.

[13]This does not assume that a vacancy meachanism is dominant; it only allows the possibility. If a vacancy mechanism does not operate, then $J_v = 0$.

To obtain Darken's flux equation, two additional assumptions must be made. These are (1) that the vacancies are everywhere in thermal equilibrium, that is $\mu_v \simeq 0$, and (2) that the off-diagonal coefficients L_{12} and L_{21} are essentially zero. Inserting these two assumptions in Eq. (4-14) gives

$$J_1 = -L_{11} \, d\mu_1/dx \qquad (4\text{-}16)$$

To relate D_1 to L_{11} in Eq. (4-16) we equate our two expressions for J_1.

$$J_1 = -L_{11} \, d\mu_1/dx = -D_1 \, dc_1/dx \qquad (4\text{-}17)$$

In Sec. 1-5, other equations were given for this flux. There a mobility was defined as the ratio of the force on an atom (F) and v the mean velocity of the atom when acted upon by F, or $M = v/F$. Using those equations, Eq. (4-17) can be rewritten as follows with the chemical potential appearing as a source[14]

$$J_1 = vc_1 = M_1 F_1 c_1 = -M_1 c_1 \, d\mu_1/dx$$

$$= -L_{11} \, d\mu_1/dx = -D_1 \, dc_1/dx \qquad (4\text{-}18)$$

It is apparent that $L_{11} = M_1 c_1$ and that

$$D_1 = M_1 \, d\mu_1/d\ln c_1 = M_1 \, d\mu_1/d\ln N_1 \qquad (4\text{-}19)$$

The second equality in Eq. (4-19) follows since if $c_1/\text{density} = N_1$ then $d\ln c_1 = d\ln N_1$. The mobility M is a more general indicator of the diffusion mobility than D as can be seen by considering spinodal decomposition. There the amplitude of fluctuations in composition increase with time, so the diffusion flux is up the concentration gradient, leading to the unphysical situation that D is negative. It can be shown that $d\mu_1/d\ln N_1 = N_1(1 - N_1)G''$ where G'' is the second derivative of the Gibbs Free Energy with respect to composition.[15] But G'' is negative for an alloy undergoing spinodal decomposition so $D_1 = M_1 N_1(1 - N_1)G''$ can go negative while the mobility M_1 remains positive and essentially constant.

The equation relating μ_1 and N_1 is

$$\mu_1 = \mu_o(T,P) + RT(\ln N_1 + \ln \gamma_1)$$

---

[14]Here the equation $F = -d\mu_1/dx$ replaces the equation $F = -dV/dx$ used in Sec. 1-5. The concentration gradient that enters $d\mu_1/dx$ does not exert mechanical force in the sense that a potential-energy gradient does, but it does produce a net flux of atoms and thus can be thought of as a force. Also one can take the viewpoint that this is required if the flux given by Eq. (4-22) is to equal zero at equilibrium.

[15]L. Darken, R. Gurry, *Physical Chemistry of Metals*, p. 240 or p. 331, McGraw-Hill, 1953.

where γ_1 is called the activity coefficient of 1. Thus

$$d\mu_1/dln\, N_1 = RT(1 + dln\, \gamma_1/dln\, N_1)$$

and

$$D_1 = M_1 RT(1 + dln\, \gamma_1/dln\, N_1) = M_1 RTI \qquad (4\text{-}20)$$

Here I is called the thermodynamic factor. The same sort of relation holds for component 2 with the same value of I. In dilute solution, γ_1 is a constant. (This follows from Raoult's Law for the solvent and from Henry's Law for the solute.) Thus in dilute or ideal solution, $D_1/M_1 RT = 1$, but in concentrated, nonideal solutions the ratio $D_1/M_1 RT$ will differ from unity. The direction and magnitude of the deviation will depend on the type and degree of nonideality. For negative deviation from ideality γ_1 is less than unity, but it increases more rapidly than linearly with N_1. Thus $dln\, \gamma_1/dln\, N_1$ is positive and D_1 is greater than $M_1 RT$. For alloys which exhibit a positive deviation from ideality D_1 is correspondingly less than $M_1 RT$. With rising temperature alloys always tend to be more ideal. Thus any deviation of $D_1/M_1 RT$ from unity will decrease with rising temperature. It will be seen in the next section that this leads to a different in the activation energy for the tracer and intrinsic diffusion coefficients.

4.4 RELATIONSHIP BETWEEN INTRINSIC D_1 AND TRACER D_1^*

The only experimental check of Darken's equations comes from a comparison of the tracer diffusion coefficients for the elements in a binary alloy and the values of D_1 and D_2 for the same alloy. Consider a binary diffusion couple with no chemical concentration gradient, that is, $dN_1/dx = 0$, but in which there is an isotopic concentration gradient, that is $dN_1^*/dx \neq 0$. (N_1^* is the mole fraction of radioactive component 1.) Using Eq. (4-20), the self-diffusion coefficient in an alloy is given by the equation

$$D_1^* = M_1^* RT(1 + dln\, \gamma_1^*/dln\, N_1^*)_{N_1+N_2} \qquad (4\text{-}21)$$

But the stable and the radioactive isotopes are chemically identical so a mixture of varying N_1/N_1^* but constant $N_1 + N_1^*$ will be an ideal solution. Since γ_1 should depend only on the overall value of $N_1 + N_1^*$ and not on the relative proportions of N_1^* and N_1, $(dln\, \gamma_1^*/dln\, N_1^*)$ in Eq. (4-21) should equal zero. Thus

$$D_1^* = M_1^* RT \qquad (4\text{-}22)$$

From this same chemical identity of the two isotopes, it seems reasonable to assume that the mobilities determined in the two types of experiments are the same, that is, $M_1^* = M_1$.[16] Substitution in Eq. (4-20) then gives

$$D_1 = D_1^*(1 + d\ln \gamma_1/d\ln N_1) = D_1^*I \qquad (4\text{-}23)$$

That is, the diffusion coefficient for component 1 in a chemical concentration gradient D_1 is not equal to the value D_1^* obtained in a self-diffusion experiment except in ideal or dilute solutions. In concentrated nonideal solutions D_1 and D_1^* will differ. This difference arises from the fact that with a given gradient dN_1/dx or dN_1^*/dx, the actual driving force, $d\mu_1/dx$, depends upon whether there is a variation in $N_1 + N_1^*$ along the diffusion direction or simply a variation of N_1/N_1^* with no gradient in $(N_1 + N_1^*)$. Note that as the temperature rises I will decrease in magnitude. Thus the activation energies for D_1 will approach that of D_1^* at higher temperatures.

The most common type of diffusion data is for $\bar{D}$, D_1^*, and D_2^*. Thus an experimental check of this analysis can be made by relating these three quantities. From the Gibbs-Duhem equation

$$N_1 d\mu_1 + N_2 d\mu_2 = 0$$

and from the definition of μ_i as a function of N_i

$$N_i d\mu_i = RT(dN_i + N_i d\ln \gamma_i)$$

Substitution of the latter into the former and the fact $dN_1 = -dN_2$ gives

$$N_1(d\ln \gamma_1/dN_1) = N_2(d\ln \gamma_2/dN_2)$$

Combining this with Eq. (4-7) and (4-23) gives

$$\bar{D} = (D_1^*N_2 + D_2^*N_1)(1 + d\ln \gamma_1/d\ln N_1) \qquad (4\text{-}24)$$

while Eq. 4-5 becomes

$$v = (D_1^* - D_2^*)(1 + d\ln \gamma_1/d\ln N_1)\partial N_1/\partial x \qquad (4\text{-}25)$$

A discussion of the fit of these equations to the experimental data is better given after a discussion of the approximations made in these equations and the first order corrections that can be made.

---

[16]Bardeen and Herring, show that for a pure metal $fM_1 = M_1^*$, where f is the correlation coefficient. Thus we are here assuming something about the relative contribution of correlation effects to experiments in which M_1 and M_1^* are determined. Our assumption is related to taking $L_{ij} = 0$ ($i \neq j$) in Eq. (4-9). This is discussed further in the next section.

4.5   TEST OF DARKEN'S ASSUMPTIONS

Two basic assumptions are implicit in the equations Darken used. First that the vacancy concentration was maintained at local equilibrium in all regions throughout the diffusion zone, and second that off-diagonal terms, L_{ij}, could be neglected.

Vacancy Equilibrium. The assumption that μ_v is zero requires that the concentration of vacancies be maintained at its equilibrium value at every point in the diffusion couple. Since diffusion occurs by a vacancy mechanism in the alloys studied, vacancy equilibrium requires that a volume of vacancies equal to the volume swept out by the markers, i.e., marker shift times cross-sectional area of the couple, must be created on one side and destroyed on the other. The problem of understanding the assumption $\mu_v \simeq 0$ is then one of determining how and where these vacancies are formed and destroyed. The regions of formation and removal can be determined by operating on the experimental $c(x)$ curve.

Figure 4-3a shows an assumed $c_A(x)$ curve for the case of $D_A + D_B$. In (b) the fluxes have been obtained by taking the gradient of $c_A(x)$ and multiplying by D_A or D_B. The vacancy flux is equal to the difference between J_A and J_B and is in the same direction as J_A. In (c) the divergence of J_v is shown, that is, the rate of vacancy generation ($dJ_v/dx > 0$) or destruction ($dJ_v/dx < 0$). From this figure it is seen that the vacancies are produced near one end of the diffusion zone and removed near the other.

Initially it was thought that the active sources and sinks of vacancies were the free surface or grain boundaries. However, it soon became clear that in metals the sources and sinks were distributed more homogeneously. Edge dislocations would serve the purpose, but must be continually regenerated since otherwise they will all grow out of the crystal. A long-lived source is obtained by having part of a dislocation with a screw component parallel to the diffusion direction rotate into an edge dislocation in a plane normal to the flux. This edge dislocation can then rotate around the screw dislocation, giving off or taking on a plane of vacancies in each revolution. This geometry is quite similar to that given by Burton, Cabrera, and Frank in their discussion of crystal growth from the vapor. In our case, their ledge is replaced by an internal edge dislocation. Bardeen and Herring[17] show that the supersaturation s required to operate a source or sink of this type is about 0.01, where

$$s = N_v(\text{actual})/N_v(\text{equilib}) - 1 \qquad (4\text{-}26)$$

[17]J. Bardeen, C. Herring, *Atom Movements*, ASM, Metals Park, OH, 1951.

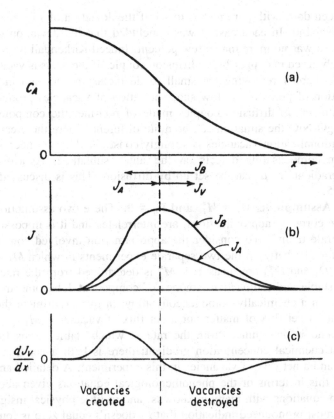

Fig. 4-3 — (a) Assumed concentration-distance curve for component A. (b) Fluxes of A and B that result ($D_B > D_A$). The flux of vacancies will be equal to the difference between J_A and J_B. (c) dJ_v/dx equals the rate of creation of vacancies at that point.

From the phenomenological viewpoint, one can say that

$$dJ_v/dx \approx - \mu_v \approx s \tag{4-27}$$

That is, the rate of creation or removal of vacancies is proportional to the deviation of μ_v from its equilibrium value of zero. Thus the line shown in Fig. 4-3c reflects the deviation from $\mu_v = 0$. This variation of μ_v with distance tends to decrease J_B and increase J_A relative to the values predicted by Darken's equations [see Eqs. (4-24) and (4-25)]. Such a deviation of μ_v from zero will therefore tend to make the observed values of D_A/D_B closer to unity than it should be. Some deviation from equilibrium is indicated by the fact that excess vacancies precipitate as voids (Kirkendall porosity) on the side of the diffusion couple where diffusion is more rapid. Several types of experiments have been done with pure metals to see if the deviation of s from zero is appreciable. In each case it was concluded that the deviation of s from zero was no more than a few percent. Indeed Kirkendall porosity is not observed in copper-brass diffusion couples if the brass is vacuum melted, thereby removing the small oxide inclusions that aid in the nucleation of pores at the low supersaturation of vacancies produced by diffusion. In diffusion couples made of non-metallic compounds like MgO-NiO the situation can be quite different. There the creation of additional cation vacancies is strongly constrained by the necessity to form compensating defects on the anion sublattice. As a result large gradients in μ_v can be set up by diffusion. This is discussed in Chap. 5.

The Assumptions $M_i = M_i^*$ and $L_{ij} = 0$. These two assumptions, or more correctly approximations, are interrelated and it is impossible to separate them.[18] To consider the approximation involved consider the difference between the two separate experiments in which M_1 and M_1^* (or D_1 and D_1^*) are measured. M_1^* is determined from the rate at which stable and radioactive isotopes of component 1 become intermingled in a chemically homogeneous alloy; in this experiment there is neither a net flux of matter nor a net flux of vacancies. M_1 on the other hand is determined from the rate at which stable atoms flow down a chemical concentration gradient; there is both a net flux of matter and a net flux of vacancies in this experiment. A detailed analysis of this in terms of the phenomenological equations given above leads to equations with many unknowns, and to little physical insight.

The most pronounced indication that L_{ij} doesn't equal zero is found by considering the effect of the net flow of vacancies through the diffusion zone. This net flux of vacancies means that in addition to their random motion, vacancies will more frequently approach any given atom from one side than from the other. This vacancy flux (sometimes called a vacancy 'wind') increases the penetration, and apparent D, for the faster moving component, and decreases the apparent D for the slower moving component. The largest experimental effect is found for the marker shift since it depends on the difference between D_1 and D_2. The corrected version of Eq. 4-25 is divided by a correlation coefficient roughly equal to that of the pure metal crystal involved.[19] For

---

[18]For a more detailed discussion of these constants, as well as experiments that indicate the degree to which they are obeyed see, T. R. Anthony, *Diffusion in Solids*, eds. A. Nowick, J. Burton, Academic Press (1975), pp. 353–79, or C. P. Flynn, loc.cit. Chap. 8.

[19]J. Manning, *Diffusion Kinetics for Atoms in Crystals*, Van Nostrand, 1968, Chap. 5. See also, C. P. Flynn, *Point Defects & Diffusion*, Clarendon-Oxford Press, (1972), Sec. 8.3.2.

an fcc lattice this correction represents an increase of about 28% in the marker shift.

$$v = (1/f)(D_1^* - D_2^*)(1 + dln\,\gamma_1/dln\,N_1)\partial N_1/\partial x \qquad (4\text{-}28)$$

Corrections to the equations relating D_1 to D_1^* are also available.

Probably the most sensitive experiments to check these equations were done by Meyer[20] on the Ag-Au system. Here the values of D_1^* and D_2^* differ by about a factor of two and the correction for non-ideality rises to 1.6. The experiments were done under an argon pressure to prevent the development of porosity. The observed marker shifts agreed with those predicted by Eq. 4-28 to within 2%. Experimental measurements on Cu-Zn[21] and Ag-Cd[22] show that a correction for the vacancy wind shifts the calculated values of D_1/D_2 appreciably and brings them into more satisfactory agreement with the observed ratio D_1^*/D_2^*.

4.6 TERNARY ALLOYS

By now the increase in complexity that can arise in going from diffusion in pure metals to binary systems should be established in the reader's mind. It is not difficult then to believe that diffusion in ternary systems is more complicated. We describe here only one experiment on ternaries. This indicates a few of the new problems that arise in going from a binary to a ternary system. It is also relevant to the problem, mentioned earlier, of the correct flux equation to be used in treating diffusion.

Consider the following experiment. A bar of an Fe-0.4%C alloy is joined to an Fe-0.4%C-4%Si alloy and diffused at 1050° C where the couple consists of only one phase (fcc austenite). Since there is no carbon concentration gradient, Fick's first law would predict no flux of carbon. Darken[23] has performed essentially this experiment and obtained the carbon distribution shown in Fig. 4-4. It is seen that carbon diffusion has increased the carbon concentration gradient in the couple, and produced a discontinuity in the concentration of carbon at the original interface.

It has been shown in other studies that adding silicon to Fe-C alloys increases the chemical potential of carbon. Also, at 1050° C the dif-

[20]R. O. Meyer, *Phys. Rev.*, *18* (1969) 1086.
[21]D. J. Schmatz, H. Domian, H. I. Aaronson, *J. Appl. Phys.*, *37* (1966) 1741-3.
[22]N. R. Iorio, M. A. Dayananda, R. E. Grace, *Met. Trans.*, *4* (1973) 1339-46.
[23]L. Darken, *Trans. AIME*, *180* (1949) 430.

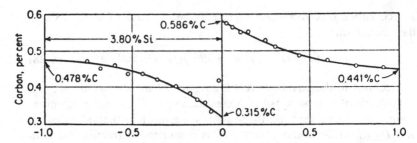

Fig. 4-4—Distribution of carbon resulting from 13-day anneal at 1050° C. The carbon content was initially 0.48% throughout the left side and 0.44% throughout the right side. (L. S. Darken) The chemical potential of carbon is continuous and monotonic decreasing across the couple throughout the diffusion anneal.

fusivity of the interstitial carbon is several orders of magnitude greater than that for the substitutional silicon. Thus upon annealing, the more rapidly diffusing carbon is redistributed in the region of the joint to give local equilibrium, that is, to eliminate the gradients in the chemical potential of carbon. Since the sudden drop in silicon concentration at the joint is not removed by diffusion, an equally sudden rise in the carbon concentration develops. Using the results of the thermodynamic studies, Darken showed that the chemical potential of carbon was indeed continuous through the joint. Thus the results can be correctly and easily described with a flux equation of the form $J = -M(\partial \mu / \partial x)$, while the description using Fick's First Law is virtually impossible.

This furnishes a striking example of the effects which can arise in ternary diffusion couples. The effect is exaggerated by the very rapid diffusion of the interstitial carbon relative to that of silicon. However, in a ternary alloy with two substitutional solutes, a flux is possible in the absence of a concentration gradient or even against a concentration gradient. In these cases the values of D obtained using Fick's laws are empirical constants which depend on the gradient of composition as well as the composition. As a result, they are of limited practical value and no theoretical value.

Before closing our discussion of Darken's experiment, it is interesting to trace out the changes in composition at two points on opposite sides of the couple using a ternary diagram. Figure 4-5 schematically shows such a trace. Note that the points initially move along lines of constant silicon concentrations. This is another way of indicating that the carbon diffuses much more rapidly than the silicon. If the silicon and the carbon diffused at the same rate, much of this curvature would be absent.

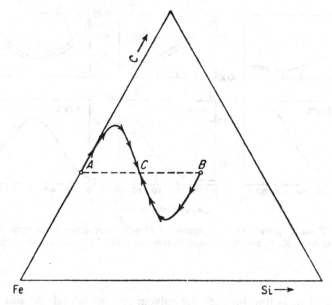

Fig. 4-5—Schematic diagram showing the change in composition with time of two points on opposite sides of the weld in Darken's diffusion couple of Fe-0.44%C and Fe-0.48%C-3.8%Si.

4.7 ESTIMATES OF $\bar{D}$ ACROSS A BINARY PHASE DIAGRAM

The metallurgist is often faced with the problem of making "reasonable approximations" for systems or alloys in which no accurate measurements have been made. In the particular case of diffusion he may be given the problem of estimating the relative value of $\bar{D}$ across much of a phase diagram which contains several intermediate phases. $\bar{D}$ has been measured in a few such alloy systems, and several helpful generalizations can be made. Two of these follow.

1. If adding A to B lowers the melting point of B, or the liquidus line, it will also increase $\bar{D}$ at any given temperature. If A raises the melting point of B, $\bar{D}$ will decrease. This rule is a variation of the observation that for a given crystal structure, $\bar{D}$ at the liquidus is roughly a constant or H/T_{liq} is roughly a constant. As an example, see Fig. 4-6.
2. For a given metal, at a given temperature and composition, diffusion will be much faster in a bcc lattice than in a close-packed

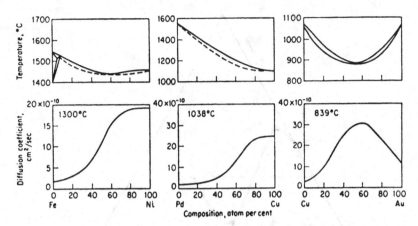

Fig. 4-6—Phase diagrams and the variation of $\bar{D}$ with composition for several iso-morphous systems. [C. E. Birchenall, *Atom Movements*, ASM, Metals Park, OH, (1951), p. 122.]

lattice. This is true for both the solvent and interstitial and substitutional solutes.

As examples of the second point, consider the following order of magnitude ratios. For carbon in essentially pure iron at 910° C, $D(\alpha)/D(\gamma) = 100$.[24] For iron in iron at 850° C, $D(\alpha)/D(\gamma) = 100$.[25] For the comparison of a bcc and an hcp lattice, in zirconium at 825° C, $D(\beta)/D(\alpha) = 10^5$. The more rapid diffusion in the bcc modification is usually "explained" by saying that the bcc lattice is more loosely packed or a more open structure. This is qualitatively consistent with our earlier discussion of the ion core repulsion to the activation energy, and indeed H_m/T_m is lower for bcc metals than for fcc metals (see Table 2-3).

PROBLEMS

4-1. Markers are placed at two locations in a diffusion couple made by welding together sheets of pure A and B so as to form two semi-infinite regions of A and B. One set of markers is placed at the A-B interface and the other a short distance away in pure A.
  • Give qualitative curves which show how the position of each marker varies with time if $D_A > D_B$. Derive these curves by

[24]R. P. Smith, *Acta Met.*, *1* (1953) 578–87. C. Wert, *Phys. Rev.*, *79* (1950) 601.
[25]C. Birchenall and R. Mehl, *Trans. AIME*, *188* (1950) 144.

plotting $N(x)$ and dN/dx versus x and using the equation

$$v = (D_A - D_B) \, dN_A/dx$$

4-2. A diffusion couple is made by joining pure Cu to Cu-10%Al, with inert markers in the weld interface. Annealing gives marker motion indicating $D_{Al} > D_{Cu}$.

(a) Draw a plot of the concentration of Al vs. distance for the annealed couple. Be sure to indicate the difference in penetration distance on the two sides of the marker interface resulting from $D_{Al} > D_{Cu}$.

(b) Show which way the markers at the interface will shift, and where the Kirkendall porosity will be growing most rapidly for the $C(x)$ curve drawn in (a).

4-3. Using Eqs. (4-22) and (4-23), calculate $dln \, D/d(1/T) = -Q/R$ for D_1 and D_1^*. Discuss the reason for, and the sign of, the difference in the two.

4-4. Consider two diffusion couples involving alpha brass:
   • a 30% Zn alloy held in a vacuum at 780° C where essentially all of the Zn that comes to the surface evaporates,
   • a piece of pure copper exposed to a Zn vapor in equilibrium with Cu-30% Zn so that the Cu surface is held at essentially 30% Zn.

(a) For each specimen, plot the variation of concentration, $C(x)$, and flux, $J(x)$, of Zn with depth below the surface. (Indicate the direction of J_{Zn} for each couple.

(b) Explain if, and where, porosity would form in each sample.

4-5. Inert markers are placed on the face of a sample of metal A, and the face is exposed to a gas which maintains a fixed concentration of B on the exposed face. The markers become embedded in the sample with time as the sample grows from the added B diffusing into the sample as A diffuses out into the B rich surface layer.

(a) Write an equation relating the difference in flux $(J_A - J_B)$ at the plane of the markers to $(D_A - D_B)$ at this plane and to the rate of marker motion relative to the pure A end of the sample. (note $N_A + N_B = 1$)

(b) Explain where the marker will end up, relative to its initial height to the external surface, if $D_B \gg D_A$, and where it will end up if $D_B \ll D_A$.

4-6. Chromium is diffused into pure iron at 1000 C under conditions which maintain a concentration of 50% Cr at the surface. The Fe-Cr system has an austenite loop; at 1000° C the maximum Cr content of the γ-phase is 12%, and the minimum Cr content of the α-phase is 13%. The diffusivity D of Cr is much greater in the α-phase than in the γ-phase. Sketch a schematic diagram of concentration of Cr vs. depth below the surface. Pay special attention to slopes and curvatures. You need not draw to scale but should indicate absolute values where possible.

4-7. In a given phase the flux of a given component is always in the direction of its decreasing chemical potential. Therefore it is helpful to establish the following. Using the thermodynamic relation between μ_1 and N_1 show that

   (a) For a binary system if $dN_1/dx = 0$, then $d\mu_1/dx = 0$, and if $d\mu_1/dx > 0$, then $dN_1/dx > 0$. (x is distance)

   (b) For a ternary system if $d\mu_1/dx > 0$ then dN_1/dx can be either >0 or <0.

4-8. Explain how D_1 and D_1^* in Darken's analysis are determined experimentally, and tell why they differ.

4-9. Consider a diffusion couple of Cu against an alpha solid solution of Cu-15a/o Al with fine inert markers initially placed in the interface between the two pieces of metal.

   (a) How would you measure the Kirkendall shift?

   (b) What is Kirkendall porosity, and where would you expect to find it?

   (c) How can this shift be related to the intrinsic diffusion coefficients D_{Al} and D_{Cu} in the alloy? (are these values for all compositions in the couple, or only for some particular composition?)

   (d) If the annihilation of vacancies was suddenly rendered impossible so that the vacancy chemical potential could locally rise, or fall, what would this do to the relative fluxes of Al and Cu?

Answers to Selected Problems

4-1. Initially dN_A/dx is large at the weld interface, and zero at the markers away from the discontinuity in composition. Thus the markers at the weld interface will start moving immediately while the other set will move only as the gradient, dN_A/dx, at that location rises.

4-3. The difference stems from the variation of the thermodynamic factor, I, with temperature. With negative deviation from ideality $I > 1$, and Q for D_1 is less than Q^* for D_1^*. With positive deviation $I < 1$ and $Q > Q^*$.

4-4. (b) Porosity could form internally in the brass losing Zn.

4-5. $dw/dt = (J_A + J_B)\Omega = (D_A - D_B)(\partial N_2/\partial x)_{x=w}$

5

DIFFUSION IN NONMETALS

In the preceding chapters, the specific examples used concerned metals. This stems partly from the author's experience but also from the fact that the majority of the research on diffusion in solids has been done with metals. There is reason to believe that all the general theory and most of the physical phenomena discussed in the earlier chapters applies equally well to nonmetals, although well-studied examples are often not available. With the change in electronic structure in going from metals to nonmetals, several new effects arise. In insulators the electrons are bound so tightly to the atoms that the principal means of carrying electric current at elevated temperature is by the movement of ions. In oxides and sulfides of transition metals the charge is carried by electrons, or electron holes, but charge conservation dictates that deviations from stoichiometry are accompanied by large increases in the concentration of the point defects that aid diffusion. In elemental semiconductors like silicon and germanium the bonding leads to both special electronic effects and the relatively easy accommodation of host and solute atoms on interstitial sites. This chapter deals with the phenomena which are unique to nonmetals, and ordered alloys.

5.1 POINT DEFECTS IN IONIC SOLIDS

The reader may recall that the forces between atoms in an ionic crystal are largely classical and that a well-developed theory of ionic crystals was established before the advent of quantum mechanics. The physical model resulting from such studies is demonstrated by the NaCl-type lattice shown in Fig. 5-1. Here the equally charged ions are arranged so that the oppositely charged ions are nearer to each other and the similarly charged ions are farther apart. If diffusion occurred by the interchange of a neighboring sodium ion and a chloride ion, these ions would go to sites which were surrounded by ions of the same charge. The increase in electrostatic energy of this new configuration over the normal situation is so great that diffusion by this mechanism is out of the question. Calculation of the energy required to form various mobile defects indicates that the dominant defects are always vacancies and interstitials.

Since there are two types of ions in these compounds, the formation of defects is not as simple as in the case of a pure metal. For example, if metal ion (cation) vacancies were formed at a surface of an ionic solid and then diffused into the crystal, the surface would have an excess negative charge; the inside of the crystal would have an equal positive charge. These separated, unlike charges would have a very large electrostatic energy per vacancy, so large in fact that the separation of unlike charges over macroscopic distances does not occur. To have defects which will maintain local charge neutrality, two kinds of defects of opposite charge must be formed. For example, if both an anion vacancy and a cation vacancy were formed, charge neutrality would be preserved. When an equal number of anion and cation vacancies are formed, the resulting disorder is said to be of the Schottky type. This type of disorder is the dominant type of disorder in alkali halide crystals and is shown in Fig. 5-1a.

An equation for the equilibrium fraction of anion sites that are vacant (N_{va}) and the equilibrium fraction of cation sites that are vacant (N_{vc})

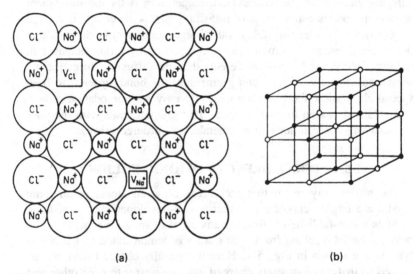

(a)   (b)

Fig. 5-1 — (a) Schematic drawing of (100) plane of NaCl lattice showing the relative sizes of the ions, and Schottky disorder: sodium vacancy V_{Na}, and chlorine vacancy V_{Cl} in equal numbers. (b) Spatial arrangement of the ions in the NaCl unit cell.

can be obtained in a manner analogous to that used in Chap. 2 to calculate N_v. If δn anion vacancies and δn cation vacancies are added to a crystal, and if all of the vacancies are randomly distributed, the change in free energy of the system will be

$$\delta G = (\delta n/N)[H_{va} + H_{vc} - T(S_{va} + S_{vc}) + RT(ln\, N_{va} + ln\, N_{vc})] \quad (5\text{-}1)$$

where $-R\, ln\, N_{va}$ is the ideal entropy increase for mixing anion vacancies into the crystal, S_{va} is the molar entropy of formation of anion vacancies, and H_{va} is the molar enthalpy of formation of anion vacancies. Similar definitions apply to $-R\, ln\, N_{vc}$, S_{vc}, and H_{vc}; N is Avogadro's number. Setting $G_{va} = H_{va} - TS_{va}$ and $G_{vc} = H_{vc} - TS_{vc}$, the condition for equilibrium is

$$(N_{va})(N_{vc}) = \exp[-(G_{va} + G_{vc})/RT] = \exp(-G_S/RT) \quad (5\text{-}2)$$

If $N_{va} = N_{vc}$ the equation can also be written $N_{vc} = \exp(-G_S/2RT)$, where G_S is the molar free energy of formation of the pair of vacancies. This defect type is found in alkali halides, e.g. NaCl. It is also found where defects on both lattices have energies of formation close enough that the motion of both defects can be measured.

If the free energy required to form an interstitial cation (G_{ic}) is much less than that required to form an anion vacancy (G_{va}), the charge of the cation vacancies will be compensated by metal ions going into interstitial sites. This combination of defects is known as Frenkel disorder. If N_{ic} is the fraction of interstitial sites which is occupied by metal ions (cations) and the defects are all randomly distributed, the equilibrium condition is

$$(N_{ic})(N_{vc}) = \exp(-G_F/RT) \quad (5\text{-}3)$$

where G_F is the molar free energy of formation for a pair of Frenkel defects, i.e., and interstitial plus a vacancy. This type of disorder is dominant in AgCl and AgBr, and is found in systems where the measurable ionic motion is essentially limited to one type of ion due to the large difference in the energies required for diffusion.

It should be emphasized that in our derivations it was not necessary to assume that $N_{vc} = N_{va}$ [in Eq. (5-2)] or that $N_{ic} = N_{vc}$ [in Eq. (5-3)]. These two equations are thus analogous to the equilibrium constants found in discussions of chemical equilibrium. Two cases commonly arise in which these general properties of the equations are used. First, if $G_{ic} \simeq G_{va}$, then $G_S \simeq G_F$; and, in addition to cation vacancies, both anion vacancies and cation interstitials will be present. The relative concentrations of the various defects must then be determined by simultaneously satisfying Eqs. (5-2) and (5-3) as well as the condition for charge neutrality which requires that $N_{vc} = N_{va} + N_{ic}$ (provided all ions have the same charge.) Second, if some of the matrix ions are replaced by ions of a different valence. For example, if $CaCl_2$ molecules are dissolved in the NaCl a proportionate number of Na^+ vacancies with a net negative charge must be added to maintain charge neutrality. Thus the equilibrium concentration of defects will be determined by the arrangement which will maintain charge neutrality and at the same time satisfy Eqs. (5-2) or (5-3).

5.2 DIFFUSION AND IONIC CONDUCTION

When a solid is placed in an electrical circuit which maintains a voltage across it, a force is exerted on the charged particles in the solid and they tend to rearrange themselves, that is, the anion and cation defects move so as to let current flow in the external circuit. In metals and semiconductors essentially all of the current is carried by electrons. However, in ionic solids at high temperatures the ions are more mobile than the tightly bound electrons. Thus electricity is conducted through the solid by the diffusion of ions.

To derive an equation relating the conductivity and the diffusion coefficient, it is necessary to take account of the force exerted on the ions by the electric field. Following the reasoning that led to the phenomenological equations of Sec. 4-4, we proceed as follows: In the absence of an electric field the condition for equilibrium is $\nabla\mu_j = 0$ for each component. If q_j is the charge on the particle then $q_j\nabla\phi$ is the force on it due to the electric field $\nabla\phi$. The condition for equilibrium in the presence of an electric field is thus

$$\nabla\phi_j + q_j\nabla\phi = 0 \tag{5-4}$$

The flux is equal to the product of the number of particles per unit volume (c_j), their mobility (M_j), and the mean force on the particles. Thus for the case in which μj and ϕ vary only along the x axis

$$J_j = -M_j c_j(\partial\mu/\partial x + q_j\partial\phi/\partial x) \tag{5-5}$$

In the absence of a field, the flux of particles can also be expressed as

$$J_j = -D_j(\partial c_j/\partial x) \tag{5-6}$$

or, from Eq. (5-5),

$$J_j = -M_j c_j \, \partial\mu/\partial x$$

Now from the definition of μ_j in dilute solutions

$$\partial\mu_j/\partial x = (RT/c_j) \, \partial c_j/\partial x$$

So these two equations for the flux will be equal if

$$M_j RT = D_j \qquad (5\text{-}7)$$

This equation is referred to as the Einstein equation or the Nernst-Einstein equation.[1]

To demonstrate how D is related to the electrical conductivity σ, consider the case in which the charge is carried by interstitial atoms. The flux of charge, or the current per unit area I, will then be given by the equation $I = J_i q_i$, and if $(\partial c_i/\partial x) = 0$, the current will be

$$I = (D_i q_i^2 c_i/RT)(-d\phi/dx)$$

The conductivity is defined by the equation $I = \sigma(-d\phi/dx)$, and the atom fraction of interstitials, N_i, equals $c_i \Omega$ where Ω is the molar volume, so we can write

$$\sigma/D_i = N_i q_i^2/\Omega RT \qquad (5\text{-}8)$$

However, using radioactive tracers, we do not measure the diffusion coefficient for the interstitials (D_i) but D_T, the diffusion coefficient for a radioactive tracer in the solid. As was shown in Chap. 2 these two are related by the equation $D_T = fD_i N_i$. Thus the conductivity is relate to the tracer diffusion coefficient by the equation.

$$\sigma/D_T = (zF)^2/f\Omega RT \qquad (5\text{-}9)$$

where the charge per ion has been replaced by the valence z times Faraday's constant to consistently work with molar units. If a vacancy mechanism of diffusion is dominant instead of an interstitial one, the same equation for the conductivity is obtained, but $D_T = fD_v N_v$ is used for D_T and the value of f would be different. Thus we have the interesting result that the relation between σ and D_T varies with the mechanism of diffusion. This will be discussed further in Sec. 5-6.

Before the equation relating σ and D_T can be used, it is necessary to know the fraction of the observed conductivity that is due to the jth type of ion σ_j. The total conductivity σ is due to the movement of anions, cations, and electrons. The fraction of the total current carried by the jth type of particle is termed the transport number t_j. Thus

$$t_a + t_c + t_e = 1$$

where t_a is the fraction of the current carried by anions, etc. It follows

---

[1] In the physics literature the term "mobility" is not applied to our M but to the product eM where e is the electronic charge. This product is often designated by the symbol μ. This procedure is not followed here since the symbol μ is used to represent the chemical potential and since the mobility as defined in Eq. (5-6) has already been used in Chap. 4.

that $\sigma_a = \sigma t_a$, etc. For alkali halide crystals at temperatures above two-thirds of their melting point, t_e is negligible. It is also found that $t_c \simeq 1$ in these compounds, although t_a can be about unity in other compounds, for example in the halides $(BaF_2, BaCl_2, PbF_2, PbCl_2)^2$ or in oxides (ZrO_2, ThO_2, CeO_2).[3] In these compounds the dominant defects giving rise to diffusion are anion Frenkel pairs or Schottky pairs. The observation that either t_a or t_c is close to unity in most ionic crystals can be understood as follows. Both D_a and D_c will vary exponentially with temperature, and the activation energies for the two will be large but different. It follows that D_a can be much larger or much less than D_c over an appreciable temperature range, so in many cases the ionic conduction will occur primarily through the movement of only one type of ion.

5.3 EXPERIMENTAL CHECK OF RELATION BETWEEN σ AND D_T

The relation between σ and D_T indicated in Eq. (5-9) can be checked experimentally and has been for several cases. The results of a study on NaCl are shown in Fig. 5-2. Here the diffusion coefficient of the sodium ion (D_T) was determined by evaporating a thin film of NaCl containing radioactive sodium on the surface of a single crystal. This was diffused, then sectioned, and D_T determined using the thin-film solution described in Sec. 1-3. In measuring the conductivity, the technique differs from that used on metals primarily in two respects. First, the currents measured are very small since σ is about 10^{-3} (ohm-cm)$^{-1}$ or 10^{-9} times that of copper at room temperature. Second, the voltage must be alternated in sign so that the crystal will not become polarized at electrodes.

The ionic conductivity of sodium can be described in terms of an apparent diffusion coefficient,

$$D_\sigma = \sigma RT\Omega/F^2 \qquad (5-10)$$

obtained from Eq. (5-9). ($z = 1$ for Na^+) In NaCl the sodium ions move by a vacancy mechanism. Since the sodium ions of NaCl form an fcc sublattice, $f = 0.78$ and Eq. (5-9) predicts that $D_T/D_\sigma = f$, so D_σ should give a line parallel to but 28% greater than the measured values of D_T. However, the results in Fig. 5-2 show that D_T essentially

---

[2]For a summary of transport number data as well as an extensive discussion of ionic conduction see A. B. Lidiard in *Handbuch der Physik*, vol. 20, p. 246, Springer-Verlag, Berlin, (1957).

[3]A. S. Nowick, in *Diffusion in Crystalline Solids*, eds., G. E. Murch, A. S. Nowick, Academic Press (1984) p. 143–88.

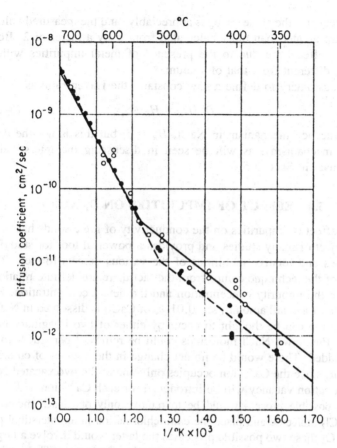

Fig. 5-2 — Log D_T vs. $1/T$ for sodium in NaCl as determined with radioactive sodium (O) and from the conductivity (●). [Mapother, Crooks, and Maurer, *J. Chem. Phys.*, **18** (1950) 1231.]

equals D_σ between the melting point and 550° C. The reason for this is unclear. One mechanism that would increase D_T without changing D_σ is the contribution of neutral pairs each being made up of a positively charged chlorine vacancy and a negatively charge sodium vacancy. The concentration of such pairs increases with the temperature for the same reason that the concentration of divacancies increases in metals (Sect. 2-9). The cation vacancies in these pairs could make a significant contribution to D_T in the high temperature domain, but because they are neutral they would not move in an electric field and thus would contribute nothing to the conductivity.[4] Below 550° C two

---

[4]V. C. Nelson, R. J. Friauf, *J. Phys. Chem. Solids*, **31** (1970) 825–43.

changes occur: the slope changes appreciably, and the measured values of D_T are greater than those calculated from σ by a factor of 2. Both of these effects are due to the presence of metal impurities with a valence different from that of sodium.

It is customary to define a new constant, the Haven ratio as

$$D_T/D_\sigma = H_R \qquad (5\text{-}11)$$

For a vacancy mechanism in NaCl, $H_R = f$, but this is not the case for all mechanisms, as will be seen in discussing the interstitialcy mechanism in Sect. 5.5.

5.4   EFFECT OF IMPURITIES ON D_T AND σ

The effect of impurities on the conductivity of ionic solids has been the subject of many studies and provides a powerful tool for studying the types and relative mobilities of the various defects formed. The power of the technique comes from the fact there are definite relations between the impurity concentration and the defect concentration. For example, if a small amount, say 0.01%, of $CaCl_2$ is dissolved in NaCl, the solution can be thought to occur by either of two imaginary processes. First, two NaCl molecules could be removed per $CaCl_2$ molecule added. There would be no net change in the number of chlorine ions but, since the Ca^{++} ion occupies only one of the two vacated Na^+ sites, a cation vacancy will be introduced for each Ca^{++} ion added. A second possible process would be to remove only one NaCl molecule per $CaCl_2$ molecule and place the extra chlorine ion in an interstitial position. Of these two possible processes the latter would involve a larger free-energy change than the former, that is, $G_{ia} > G_{vc}$. Thus, when $CaCl_2$ is added to NaCl, a cation vacancy will be introduced for each Ca^{++} ion. To maintain charge neutrality, the number of defects with a net positive charge (Ca^{++} ions and anion vacancies) must equal the number of defects with a net negative charge. The relation between the atom fraction of the impurities and the atom fraction of defects is therefore

$$N^{++} + N_{va} = N_{vc} \qquad (5\text{-}12)$$

where N^{++} is the fraction of cation sites occupied by divalent impurities.

The conductivity and diffusivity are both proportional to the concentrations of mobile defects so that the problem is to see how this concentration is changed by the impurity concentration. We discuss the case of Schottky disorder in the NaCl-type lattice. Frenkel disorder is treated in the next section. In a pure NaCl-type crystal the concentration of anion vacancies (N_{va}^0) must equal that of the cation vacancies

(N_{vc}^0). (The superscript zero is used to denote the pure material.) If divalent impurities are added, the concentration of cation vacancies and anion vacancies will be different, but Eq. (5-2) is still valid. Thus using Eq. (5-12) to substitute for N_{va} gives

$$(N_{vc})(N_{vc} - N^{++}) = \exp(-G_S/RT) = (N_{vc}^0)^2 = (N_{va}^0)^2 \quad (5\text{-}13)$$

This equation assumes that there is no significant interaction between the impurities and the defects, so they are randomly distributed, i.e., the solution is ideal. Equation (5-13) can be rewritten

$$\left(\frac{N_{vc}}{N^{++}}\right)^2 - \frac{N_{vc}}{N^{++}} - \left(\frac{N_{vc}^0}{N^{++}}\right)^2 = 0$$

This is a quadratic equation in (N_{vc}/N^{++}). Now N_{vc}/N^{++} must be positive, so the root of the equation is

$$N_{vc} = \frac{N^{++}}{2}\left[1 + \left(1 + \frac{(2N_{vc}^0)^2}{N^{++2}}\right)^{1/2}\right] \quad (5\text{-}14)$$

This equation simplifies in two limiting cases—first, when $N_{vc}^0 \gg N^{++}$, N_{vc} approaches N_{vc}^0; second, when $N_{vc}^0 \ll N^{++}$, N_{vc} approaches N^{++}. The first case applies to pure material, while the second applies to impure material. However, N_{vc}^0 varies exponentially with temperature, and since no material is absolutely free from multivalent impurities, there will always be a temperature below which $N_{vc}^0 \ll N^{++}$.

We can now explain the sharp change in slope at 550° C shown in Fig. 5-2. If sodium diffuses by a vacancy mechanism, D_T will be given by an equation of the form

$$D_T = \gamma \, a^2 \, N_{vc} \, w_T$$

where γ is a constant roughly equal to 1, and w_T is the jump frequency of a sodium tracer next to a vacant site. In the temperature range where $N^{++} \ll N_{vc}^0$, this equation can be written

$$D_T = \gamma a^2 \exp(-G_S/2RT) \exp(-G_m/RT)$$

$$= D_o \exp[-(H_S/2 + H_m)/RT] \quad (5\text{-}15)$$

The variation of D_T with temperature here stems from the fact that both N_{vc} and w_T vary with temperature. The observed D_T, and thus σ, are independent of the purity or history of the specimen in this range so that these properties are intrinsic to the pure compound; the name "intrinsic range" was applied to the phenomenon before the mechanism was understood.

In the low-temperature range $N^{++} \gg N_{vc}^0$. Here the vacancy concentration no longer varies with temperature but equals N^{++}. The equation for D_T is then

$$D_T = D_o' \exp(-H_m/RT) \qquad (5\text{-}16)$$

where D_o' is much less than D_o and is roughly equal to $N^{++}D_o$. This behavior is called "extrinsic" since it depends on the impurity content rather than the intrinsic properties of the crystal.

The validity of Eqs. (5-15) and (5-16) can be checked in at least two ways. First, the difference between the observed slope in Fig. 5-2 in the intrinsic range and that in the low-temperature or extrinsic range should equal $H_S/2$. The measured activation energies for the two ranges is shown in Table 5-1.[5] The indicated value of H_S is 2.06 eV. This agrees well with the calculated value of 1.92 eV.[6] The values of the ratios of D_o to D_o' are also given. The crystals used were reported to be better than 99.99% pure; this agrees with the observed value of D_o/D_o' and the prediction $D_o/D_o' \simeq 1/N^{++}$.

Table 5-1. Diffusion Data for Sodium in NaCl[a]

Intrinsic Range	
$D_o = 3.1$ cm^2/s	$H_m + H_S/2 = 1.80$ eV
Extrinsic Range	
$D_o' = 1.6 \times 10^{-6}$ cm^2/s	$H_m = 0.77$ eV
$D_o/D_o' = 2 \times 10^6$	$H_S = 2.06$ eV

[a]D. Mapother, H. Crook, R. Maurer, *J. Chem. Phys.*, *18* (1950) 1231.

The other check on this model comes from measuring the conductivity of a series of crystals which contain controlled amounts of divalent impurities. If the model leading to Eq. (5-16) is correct, the values of σ, or D_T, should increase linearly with N^{++} at any given temperature in the extrinsic range. Experiment shows that this is essentially the case. Values of σ for NaCl doped with CdCl$_2$ are shown in Fig. 5-3. The values of D_o' increase with N^{++}, while the slope is unchanged. Also, when the temperature is high enough that the extrinsic conductivity of each crystal becomes less than that of pure NaCl, the slope increases, and the intrinsic conductivity becomes dominant.

If σ of the above study is plotted against N^{++}, that is N_{Cd}, at some fixed temperature in the extrinsic range, it is found that the line is not quite straight but is concave downward. Thus σ does not quite increase linearly as expected. The number of vacancies must still be equal to

[5]For a more complete set of formation energies see L. W. Barr, A. B. Lidiard, in *Physical Chem.*, v. 10, ed. W. Jost, Academic Press (1970) 151–228.
[6]M. Tosi and F. Fumi, quoted by Lidiard, op. cit., p. 274.

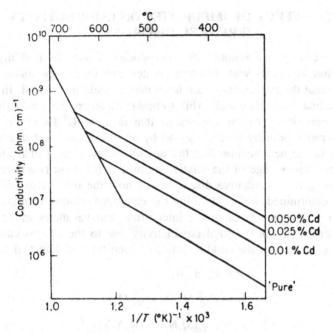

Fig. 5-3—Log σ versus $1/T$ for NaCl crystals doped with varying atom percent of $CdCl_2$.

N^{++} so all of the vacancies must not be contributing to σ. This ineffectiveness of some vacancies stems from the fact that there is a coulombic attraction between the divalent impurities and cation vacancies. Thus there will be an equilibrium of the form

$$V_{Na} + Cd^{++} = V\text{–}Cd$$

where $V\text{–}Cd$ designates a cation vacancy on a site next to a cadmium ion and bound to it. As the temperature drops the thermal energy drops relative to this binding energy and more of the sodium vacancies become bound to Cd^{++} ions. Since these pairs are neutral they will not migrate toward the anode as an unassociated vacancy would, and thus they do not contribute to σ. However, the bound vacancy can contribute to the motion of tracer atoms, and thus D_T, almost as effectively as a free vacancy.[7] This binding to form neutral pairs is probably the main cause of D_T/D_σ being roughly two in Fig. 5-2. This binding will become more effective as the temperature drops further so there is no fixed value of the Haven ratio in this regime.

---

[7] For a calculation of σ/D_T in this case see Lidiard, op. cit., p. 332.

5.5 EFFECT OF IMPURITIES ON CONDUCTIVITY (FRENKEL DISORDER)

In the example of Schottky disorder studied above, the mobility of the cation vacancies was so much greater than that of the anion vacancies that the movement of the latter defect could be ignored. In the best-studied materials which exhibit Frenkel disorder, the conductivity occurs entirely by cation movement, that is, $t_c = 1$.[8] However, the cations move both by a vacancy and by an interstitial mechanism. It is seen in the next section that the unambiguous relation of D_T to σ requires a knowledge of the relative mobilities for these two defects. Here we give a qualitative discussion of how the relative mobilities can be determined, before deriving the equations relating σ to D_T.

To a good approximation the interstitials and vacancies move independently so that the total conductivity due to the cations can be taken as the sum of the conductivities of each type of defect; thus

$$\sigma = \sigma_i + \sigma_v$$

$$\sigma = c_i q_i^2 M_i + c_v q_v^2 M_v \qquad (5\text{-}17)$$

$$\sigma = c_c q_c^2 (M_i N_i + M_v N_v) \qquad (5\text{-}18)$$

where c_c is the number of cations per unit volume, and we have used the relations $c_i = c_c N_i$, $c_v = c_c N_v$ and $q_c = q_i = -q_v$.[9]

If the ratio N_i/N_v is varied without changing the ratio M_i/M_v, the change of σ with N_i/N_v will indicate whether or not the defect being added is the more mobile. This can be done as follows: if small amounts of $CdBr_2$ are added to $AgBr$, the various concentrations (atom fractions) must obey the equations—from charge neutrality:

$$N^{++} + N_{ic} = N_{vc}$$

from equilibrium:

$$(N_{vc})(N_{ic}) = (N_{vc}^0)^2 = (N_{ic}^0)^2 = \exp(-G_F/RT) \qquad (5\text{-}3)$$

Thus as the concentration of divalent ions is increased the concentration of interstitials will decrease, and that of the vacancies will increase. Experimentally, it is found that the conductivity decreases when $CdBr_2$ is first added to pure $AgBr$. This shows that silver interstitials are more mobile than silver vacancies.

---

[8]Oxides of MO_2 structure can have anion interstitial and vacancies as the main defects. They are also well studied, see A. S. Nowick, in *Diffusion in Crystalline Solids*, ed. G. E. Murch, A. S. Nowick, Academic Press, 1984, p. 143–188.

[9]The equations developed here are for the case of a compound made up of monovalent ions, for example AgBr.

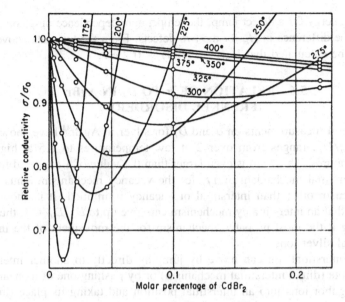

Fig. 5-4—Relative conductivity σ/σ_o of AgBr vs. atom fraction of CdBr$_2$ at several temperatures. [After J. Teltow, *Ann. Physik*, 5 (1949) 63.]

As the concentration of CdBr$_2$ is increased, σ continues to decrease as long as the majority of the conduction is due to interstitials. However, as N^{++} and N_{vc} continue to increase, there comes a composition at which σ goes through a minimum and starts to increase (see Fig. 5-4). In this range the vacancies are starting to give the majority of the conduction. Using the concept developed above, an equation can be obtained which expresses σ/σ_o in terms of N^{++}, N_{vc}^0, and M_i/M_v.[10] (σ_o is σ at N^{++} = 0). Measurements of σ/σ_o versus N^{++} at various temperatures then allow the determination of M_i/M_v and N_{vc}^0 from the values of σ/σ_o and N^{++} at the minima. The equations are

$$(\text{N}^{++})_{\min} = N_{vc}^0(\phi - 1)/\sqrt{\phi} \qquad (5\text{-}19)$$

$$(\sigma/\sigma_o)_{\min} = 2\sqrt{\phi}/(1 + \phi) \qquad (5\text{-}20)$$

where $\phi = M_i/M_v$. Figure 5-4 shows that $(\sigma/\sigma_o)_{\min}$ increases as the temperature increases. In view of Eq. (5-20), this means that ϕ decreases with increasing temperature. Teltow found that for AgBr, ϕ changes from 7 as 180° C to 2 at 350° C. Since the temperature dependence of each defect mobility is determined entirely by the acti-

---

[10]J. Teltow, *Ann. Physik. Leipzig*, 5 (1949) 63. (in German). For a discussion in English, see Lidiard, op. cit. p. 288.

vation energy for a defect jump, the temperature dependence of ϕ comes from the difference in H_m for the two defects. From the two data above, it can be calculated that $H_m(v) - H_m(i) = 0.2$ eV.

5.6   RELATION OF σ TO D_T IN AgBr (FRENKEL DISORDER)

Careful measurements on σ and D_T for silver in AgBr[11] have shown that D_σ/D_T, ranges from over 2 at low temperatures to 1.5 at high temperatures. This ratio is even larger than the value of 1 expected for an interstitial mechanism or $1/f$ for the vacancy mechanism. Thus a mechanism other than interstitial or vacancy is implied. It is shown below that an interstitialcy mechanism can give up to $D_\sigma/D_T = 3$, thus making it the most probable mechanism for the movement of the interstitial silver ions.

An interstitial ion can move by jumping directly to another interstitial site (direct interstitial mechanism) or by pushing one of its nearest-neighbor ions into an interstitial position and taking its place (interstitialcy mechanism). If the latter occurs, the interstitial charge moves farther than either of the two ions involved. Figure 5-5 shows an interstitial ion in one of the eight interstitial positions in an AgBr unit cell. A heavy arrow indicates a jump to its nearest neighbor in the exact center of the cell. This center atom can be displaced in a forward direction to the center of any one of four cubes. If it goes in the direction of the heavy arrow, [111], the jump is called a collinear jump. If it goes to one of the other three forward positions, it will be called a noncollinear jump.

To relate D_T to σ for the interstitialcy mechanism, we again start with Eq. (5-8)

$$\sigma/D_i = N_i(zF)^2/\Omega RT \qquad (5-8)$$

where D_i is the diffusion coefficient for the interstitial ion. The cation sites neighboring an interstitial cation are all equivalent, so that all jump directions are equally probable and successive jump directions of the interstitial are uncorrelated. Thus

$$D_i = (1/6)\Gamma_i \alpha_i^2$$

It should be emphasized that this equation refers to the diffusion of the interstitial. To clarify the relation between the movement of the interstitial and the movement of a particular ion, consider the two-dimensional square lattice shown in Fig. 5-6. Between Fig. 5-6a and

[11]R. Friauf, *Phys. Rev.*, *105* (1957) 843. For similar work on AgCl see M. D. Weber, R. J. Friauf, *J. Phys. Chem. Solids*, *30* (1969) 407–19.

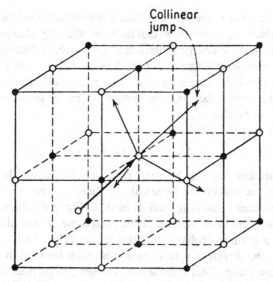

Fig. 5-5—The heavy arrow shows the movement of the interstitial atom making a jump by the interstitialcy mechanism. The other four arrows represent the possible jumps of the atom displaced from a normal lattice position. The collinear jump vector is so labeled; the three possible non-collinear jump vectors are not.

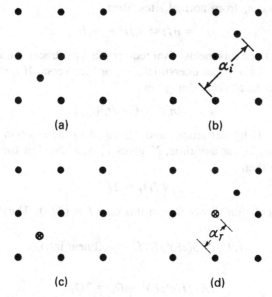

Fig. 5-6—(a) and (b) show schematic diagrams of a two-dimensional square lattice before and after a collinear interstitial jumps has occurred. α_i is the distance which the interstitial moves in this process. (c) and (d) show the same process, except that a tracer atom (0) is involved. α_T the distance which the tracer moves in this process.

b the interstitial has moved by a collinear interstitialcy mechanism. As far as a conductivity experiment can indicate, the only change between (a) and (b) is that the charged interstitial has moved a distance α_i. The same process is shown again in Fig. 5-6c and d, except that here a tracer is involved. In a tracer experiment a particular ion is followed instead of the interstitial, and the distance a tracer ion moves in the jump α_T is one-half of α_i.

The equation for D_T is

$$D_T = (1/6) f \Gamma_T \alpha_T^2$$

$f < 1$ in this case as can be seen from the following. In 5-6c the neighbors of the tracer are identical, so there can be no correlation between the tracer's last jump and its next. In Fig. 5-6d the tracer must make its next jump to an interstitial site, and these are not all identical. Thus in (d) it is more probable that the tracer will make its next jump backward in the direction it just came from than forward in the direction of its last jump. This correlation between jumps makes $f < 1$.

To derive an equation relating D_i to D_T for the collinear interstitialcy mechanism, we note first that $2\alpha_T = \alpha_i$ so that $D_i/D_T = \Gamma_i/(4 f \Gamma_T)$. The relation between Γ_T and Γ_i can be obtained as follows: If over a long period of time t a tracer ion makes n jumps, n_i of which are interstitial and n_n from normal sites, then

$$\Gamma_T = n/t = n_i/t + n_n/t$$

but $n_i = n_n$, since this mechanism requires that the tracer always jumps from a normal site to an interstitial site, or vice versa. If t_i is the time spent in interstitial sites, this gives

$$\Gamma_T = 2n_i/t = (2n_i/t_i)(t_i/t)$$

Now $n_i/t_i = \Gamma_i$ by definition, and t_i/t equal to the fraction of atoms on interstitial sites at any time, N_i gives $\Gamma_T = 2\Gamma_i N_i$. For the collinear interstitialcy then

$$D_i N_i/D_T = 2/f \qquad (5\text{-}21)$$

Detailed calculation shows that in this case $f = (2/3)$. Thus with Eq. (5-8)

$$\sigma/D_T = 3(zF)^2/RT\Omega \quad \text{(collinear int.)}$$

or

$$\sigma RT\Omega/c(zF)^2 = D_\sigma = 3D_T \qquad (5\text{-}22)$$

For a noncollinear mechanism, again $2\,\Gamma_i N_i = \Gamma_T$, but now $(a_i/a_T)^2 = (8/3)$ instead of 4. Also the value of f is lower in this case. The result is

$$D_\sigma = 1.38 \, D_T \qquad (5\text{-}23)$$

The third mechanism that contributes to diffusion in AgBr is a vacancy mechanism. This was discussed above. Since $\alpha_T = \alpha_v$, the entire effect is due to correlation, and from Eq. (5-9)

$$D_\sigma = 1.27 \, D_T \quad \text{(vacancy)} \qquad (5\text{-}24)$$

Experimental results for AgBr are shown in Fig. 5-7. The ratio D_σ / D_T varies from 2.17 (140° C) to 1.50 (350° C). It is apparent from the

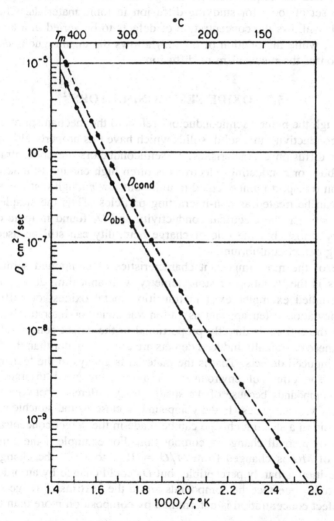

Fig. 5-7—Log D versus $1/T$ for silver in AgBr as determined by tracer and conductivity experiments. [From R. J. Friauf, *Phys. Rev.*, *105* (1957) 843.]

large value of D_σ/D_T that the collinear interstitialcy mechanism must account for an appreciable part of the conductivity at 140° C, and for a smaller, but finite, fraction at the higher temperature. With the help of Teltow's results on the relative mobilities of vacancies and interstitials in AgBr (see Sec. 5-5), Friauf was able to determine the relative contributions of collinear and noncollinear jumps as a function of temperature.

In the last two sections, we have shown that the relation between σ and D_T and the requirement of charge neutrality combine to give a unique set of tools for studying diffusion in ionic materials. Charge neutrality allows the concentration of defects to be varied in a known manner, while the relation of σ to D_T allows the unambiguous determination of the mechanism of diffusion.

5.7   OXIDE SEMICONDUCTORS[12]

Though the name "semiconductor" refers to the mechanism of electrical conductivity in a solid, solids which have this property also have similar diffusion characteristics. In semiconductors the concentration of mobile, or conducting, electrons is often high enough to make the electron transport number equal to unity but low enough that the electrons can be treated as non-interacting particles. Thus the simple relation between the electrical conductivity and D_T found in ionic conductors is lost, but the rule of charge neutrality can still be used in treating defect equilibrium.

One of the most important characteristics of compound semiconductors is the variation of stoichiometry with annealing atmosphere. Well studied examples exist in transition metal oxides, or sulfides. This deviation often appears as cation vacancies or interstitials, and D_T for the cations is directly proportional to the defect concentration, since these chemically induced defects are always greater than the thermally induced defects. That is the material is always in the 'extrinsic' range. The effect of environment is more pronounced in transition metal compounds because of the small energy difference between their different valence states. If the compound is close to the stoichiometric composition a striking change can be made in the defect concentration with only a small change in composition. For example if the composition of MO is changed from $M/O = 10^{-5}$ to 10^{-4}, the change in composition is barely perceptible, but D_T could change by an order of magnitude. In effect the compounds are in the "extrinsic" range since the defect concentration is determined by composition more than ther-

---

[12]For oxide diffusion data, see R. Freer, *J. Matl. Sci.*, *15* (1980) 803–24.

mal fluctuations. The study of this type of behavior is important in understanding the kinetics of high temperature alloy oxidation.[13]

CoO. As an example of this type of behavior and the information obtainable therefrom, consider the case of CoO. In this compound the atomic radius of the oxygen ion is twice that of the cobalt so diffusion occurs only on the cation sublattice. The variation of the vacancy concentration with oxygen partial pressure can be obtained by an equation of the form

$$1/2\ O_2(g)\ +\ 3\ Co_{Co}^{2+} = CoO\ +\ 2\ Co_{Co}^{3+} + V_{Co} \qquad (5\text{-}28)$$

where V_{Co} designates a cobalt ion vacancy. The cobalt ions are given the subscript 'Co' to emphasize that sites are conserved in the reaction as well as charge, etc. The electron deficit which converts a Co^{2+} ion into a Co^{3+} ion is called an electron hole (h^+), and is essentially an electron vacancy. At any instant it is associated with one ion, but it can easily move from one to another.

The equilibrium constant for the reaction shown in Eq. (5-28) is

$$K = [a_{CoO}(a_{Co3+})^2 a_V]/[(a_{Co2+})^3\ (P_O)^{1/2}] \qquad (5\text{-}29)$$

CoO is present as a pure solid so CoO and Co^{2+} will have activity of one. The holes, vacancies, and oxygen will be present at low concentrations. If it is assumed that none of these species interact, i.e., the solution is ideal, the equilibrium constant for the reaction is

$$(N_h^2)(N_v)/(P_O)^{1/2} = b^3\ \exp(-H'/RT) \qquad (5\text{-}30)$$

where b is a constant and H' is the molar enthalpy change for the reaction. Since two Co^{2+} ions are oxidized to Co^{3+} to form each cation vacancy, charge neutrality requires that $2N_v = N_h$. Inserting this relation in Eq. (5-30) gives

$$N_v = (P_O)^{1/6} b\ \exp(-H'/3RT) \qquad (5\text{-}31)$$

If this equation is substituted in the expression $D_T = \gamma a_o^2 N_v w_T$, two results follow. First, the diffusion coefficient at constant temperature should increase as the 1/6th power of the oxygen partial pressure. Second, the quantity H in the expression

$$\partial \ln D_T/\partial(1/T)_{P_O} = -H/R$$

will not equal the heat of motion H_m but will equal $H_m + H'/3$. This reflects the fact that at constant oxygen pressure the composition of

[13]See for example N. Birks, G. H. Meier, *Intro. to High Temp. Oxidation of Metals*, Edward Arnold (1983)., and H. Schmalzried, *Ber. Bunsenges. Phys. Chem.*, 87 (1983) 551–8, 88 (1984) 1186–94.

the oxide changes with temperature. Note that an equation essentially identical to Eq. (5-31) would hold for the concentration of electron holes. The electrical conduction in CoO is entirely by holes, so the measurement of the electrical conductivity as a function of oxygen partial pressure and temperature gives information on how D_T will change with these variables.

Experimental results for D_{Co} are shown in Fig. 5-8.[14] The diffusion coefficient clearly increases with oxygen pressure, rising with about a slope of 1/5th on the low oxygen side of the stability range, and 1/4th on the high side of the range. The electrical conductivity shows exactly the same pressure and temperature dependence. There are sev-

---

[14]The discussion of CoO largely follows R. Dieckmann, Z. Phys. Chem. Neue Folge, 107 (1977) 189–210.

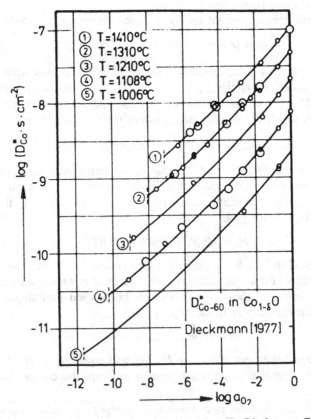

Fig. 5-8—Tracer D of Co in CoO vs. oxygen activity. [R. Dieckmann, Z. Physik. Chem. NF, 107 (1977) 189.]

eial possible explanations for this rise, which is a deviation from ideality due to the 'large' concentration of vacancies and holes.

One model to account for this coulombic attraction between the positively charged electron holes and the cation vacancies (which have an effective charge of -2) is as follows. Due to this attraction, the holes and vacancies will not be randomly mixed, but there will be a high probability of a vacancy having at least one electron hole on one of its nearest-neighbor sites at any instant. If this situation is approximated by the case in which half of the electron holes are randomly distributed and the other half are bound to vacancies to form vacancy-hole complexes $(V_c h^+)$, the reaction $Co_{Co}^{2+} + V_c h^+ = Co_{Co}^{3+} + V_c$ should be subtracted from Eq. (5-28) giving

$$1/2\ O_2(g)\ +\ 2\ Co_{Co}^{2+} = CoO + Co_{Co}^{3+} + V_c h^+ \qquad (5\text{-}32)$$

But charge neutrality requires that the concentration of holes and complexes are equal so

$$(N_h)(N_{vh})/(P_O)^{1/2} = K(T)$$

It follows that for this case N_v and $N_v h$ are proportional to $P_O^{1/4}$.

If not half but all of the holes are bound to vacancies, a similar analysis shows that the concentration of such defects $(h^+ V_c h^+)$ would vary as the square-root of the pressure. The experimental results for CoO would indicate that each vacancy is partially paired with roughly one hole, at the higher oxygen potentials and less than one at lower oxygen pressures.

Magnetite (Spinel). A second interesting example of defects in a transition metal oxide is found in Fe_3O_4 (magnetite). This is a spinel structure containing divalent and trivalent cations in the ratio $1:2$ $(Fe_3O_4) = ([Fe_2O_3][FeO])$. It is metal deficient relative to the stoichiometric composition at all but the lowest oxygen partial pressures for which the phase is stable. If D_T for iron is measured over the range of oxygen partial pressures allowable in magnetite, the results shown in Fig. 5-9 are found.[15] At high oxygen activities D_T increases as $(P_O)^{2/3}$, as one might expect for a vacancy mechanism from the discussion of D_T in CoO given above. However, as P_O decreases, D_T goes through a minimum and then rises in a manner reminiscent of the results for interstitial diffusion in AgBr shown in Fig. 5-4.

The defect structure developed in Fe_3O_4 with changing P_O in the range on the right of Fig. 5-9 can be described by the equation

[15]R. Dieckmann, H. Schmalzried, *Ber. Bunsenges. Phys. Chem.*, *81* (1977) (I)344–7, (II)414–19.

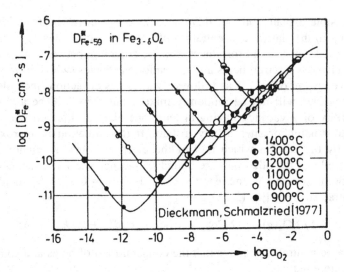

Fig. 5-9—Tracer D of Fe in Fe_3O_4 vs. oxygen activity. (R. Dieckmann, and H. Schmalzried.)

$$2/3 \, O_2(g) + 3 \, Fe_{Fe}^{2+} = 1/3 \, Fe_3O_4 + 2 \, Fe_{Fe}^{3+} + V_{Fe} \qquad (5\text{-}33)$$

where the subscript 'Fe' is added to V and Fe so that one can see that atomic sites are conserved in the reaction as well as chemical elements and charge. As the oxygen activity drops, the compound shifts toward stoichiometry (the concentration of oxygen vacancies drops). The activity is unity for Fe^{2+}, Fe^{3+}, and Fe_3O_4 so Eq. (5-33) indicates that the vacancy concentration will vary as $a_O^{2/3}$ in this regime. However, as the oxygen activity drops the oxide becomes so close to stoichiometric that the interstitial-vacancy concentration is determined by the reaction $Fe_{Fe}^{2+} = Fe_I^{2+} + V$ which corresponds to the formation of Frenkel pairs. Adding this to Eq. (5-33) gives

$$2/3 \, O_2(g) + 2 \, Fe_{Fe}^{2+} + Fe_I^{2+} = 1/3 \, Fe_3O_4 + 2 \, Fe_{Fe}^{3+} \qquad (5\text{-}34)$$

Thus near stoichiometry the concentration of iron interstitials varies as $(a_o)^{-2/3}$. The model then predicts

$$D_T = aD_v(P_O)^{2/3} + bD_I(P_O)^{-2/3} \qquad (5\text{-}35)$$

as is observed. (a and b are constants) The minimum in D_T occurs when the two terms in Eq. (5-35) are equal, while the stoichiometric composition would occur when the concentrations of cation interstitials and vacancies are the same. A careful comparison of the stoichiometric composition with the diffusion data shows that the minimum in D_T comes at a higher value of a_O than does stoichiometry. It is calculated that $D_I =$ 15 D_V at this composition.

5.8 ELEMENTAL SEMICONDUCTORS[16]

Diffusion and its control plays a major role in the production of transistors and other solid state devices, since the performance of the device depends critically on the distribution of dopants. The atomic bonds in semiconductors are covalent (homopolar) rather than metallic. This results in several basic differences between Si and Ge on one hand and say metallic Cu or Ni on the other. For example:

• the diamond cubic lattices of Ge and Si are much more open than that of close-packed metals. That is, there is more space between the hard ion cores.

• the formation energy of vacancies is higher, relative to the melting temperature, and the formation energies of vacancies and self-interstitials are more nearly equal than in metals.

• atoms much more often occupy interstitial positions in Si and Ge. This leads to an important role for Si interstitials in self diffusion, and to relatively large atoms (Fe, Ni, and Cu in Si) occupying both lattice and interstitial positions at equilibrium. Motion of the interstitials dominates the diffusion process.

• the presence or absence of bonding can determine the mobility of the solute. For example oxygen is small and occupies interstitial sites in Si but bonds with Si and diffuses with a relatively high activation energy (2 eV) while the larger Ni and Cu atoms, which form no bonds with Si, move with a Q of closer to 0.5 eV.

There are good isotopes for the determination of self diffusion in germanium, but no appropriate isotope for silicon. Thus D for Ge is well known while that for Si is less certain. However, it is clear that self-diffusion in Ge and Si at their respective melting temperatures is orders of magnitude slower than that in metals at their melting points. (see Fig. 5-10) This difference between semiconductor and metal increases yet more at lower temperatures due to the relatively larger activation energies (Q/T_m) for Ge and Si (see Table 2-5). A variety of experiments indicate that self diffusion in Ge and Si is dominated by vacancy motion at low temperatures, but at high temperatures D in silicon may be dominated by the motion of interstitials. The values of

[16]W. Frank, U. Goesele, H. Mehrer, A. Seeger, *Diffusion in Crystalline Solids*, eds. G. E. Murch, A. S. Nowick, Academic Press (1984), p. 64–142. S. M. Hu in *Atomic Diffusion in Semiconductors*, ed. D. Shaw, Plenum (1973), pp. 217–350.

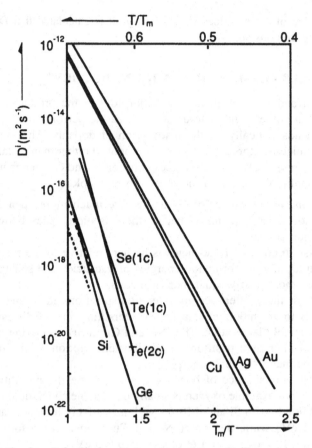

Fig. 5-10—Log D for Si, Ge and the noble metals vs. T_m/T.

D_o for self diffusion in Si and Ge are in the same range as that for metals. (See Table 2-5)

Fig. 5-11 plots log D vs $1/T$ for a variety of solutes in Si. The lines representing D for Group III and V elements lie above but roughly parallel to those for Si self diffusion. In addition there are some solutes that diffuse much faster, and with lower activation energies, e.g. Cu, Ni, Au or Li. These fast diffusing metal atoms are believed to form a complete electron shell by capture of an electron or electron hole. The resulting atoms interact weakly with the matrix and as a result the migration energy is low (<1 eV). The fast diffusing solutes reside both on substitutional and interstitial sites. Depending on the solute, the atoms may occupy predominantly substitutional or predominantly interstitial sites at equilibrium. However, there will always be a distribution of atoms between the two types of sites, and the much higher

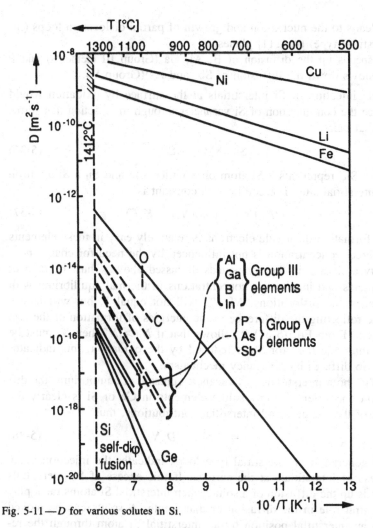

Fig. 5-11—*D* for various solutes in Si.

mobility of interstitial atoms means that they usually make a dominant contribution to the diffusion of such solutes. In fact the main new theme of this section is the role of interstitials in self and solute diffusion and the factors that influence the local concentration of interstitials.

The mechanism of diffusion in elemental semiconductors has been deduced from several types of experiments. Two are discussed briefly here.

Influence of Oxidation on Diffusion. In the manufacture of silicon based devices it is customary to oxidize the surface at intermediate temperatures to form an insulating SiO_2 layer. It has been found that the growth of this oxide layer has three effects:

• it leads to the nucleation and growth of partial dislocation loops (interstitial type) on $\{111\}$ planes.
• it speeds up the diffusion of B, Al, Ga (Group III elements) and P
• it slows down the diffusion of Sb, and As (Group V).

The injection of Si interstitials at the surface by oxidation would reduce the concentration of Si vacancies through the reaction that forms Frenkel pairs.

$$Si_I + V_{Si} = Si_{Si} \qquad (5\text{-}36)$$

where Si_{Si} represents a Si atom on a lattice site and Si_I a Si atom on an interstitial site. The equilibrium constant is

$$(N_I)(N_v) = (N_I^e)(N_v^e) = K(T) \qquad (5\text{-}37)$$

The formation of mobile electrons is relatively easy in these elements so defect concentration is not influenced by the need for charge neutrality as it was in the compounds discussed above. Thus the ratio of vacancies and interstitials concentrations in thermal equilibrium with surfaces and dislocations, N_I^e/N_v^e, will not equal 1, but will take a value reflecting the difference in the energies of formation of the two defects. From Eq. (5-37) it follows that if $N_I > N_I^e$ then N_v must be less than N_v^e. If D for Sb is reduced by the oxidation, this indicates that Sb diffuses by a vacancy mechanism.

With both interstitials and vacancies present at equilibrium, the diffusion coefficient for any solute elements in Ge or Si is clearly the sum of the vacancy and interstitial contributions, thus

$$D = D_I N_I + D_v N_v \qquad (5\text{-}38)$$

The generation of interstitial type loops indicates the injection of an excess of interstitials at the surface. If this excess of Si interstitials speeds up the diffusion of a solute, then interstitial Si atoms must play a central role in the diffusion of that solute. A solute can be moved into an interstitial position by an interstitial Si atom through the reaction

$$A_I = A_{Si} + Si_I \qquad K = N_s N_I / N_i \qquad (5\text{-}39)$$

This is termed the 'kick-out' mechanism, whereby an interstitial silicon pushes an interstitial solute off of a normal lattice site and into an interstitial position.

Since an excess of Si interstitials would lead to an increase in the concentration of interstitial solute atoms, and is found to speed up the diffusion of Group III elements (B, Al, Ga), and P, it is clear that the diffusion of these solutes is determined primarily by their motion as interstitials. This could occur if the solute diffused either by an interstitial or an interstitialcy mechanism. It appears that the dominant mechanism is by an interstitialcy mechanism rather than by simple interstitial motion. For Group V solutes the mechanism depends on the solute radius. Phosphorous, the smallest atom in this group, diffuses primarily by an interstitial mechanism while the largest, Sb, diffuses primarily by a vacancy mechanism.

Fast Diffusers. Figure 5-11 indicates that some of the transition and noble metals diffuse many orders of magnitude faster than the host silicon atoms. Similar results have been reported for diffusion in germanium. The cause for this is two-fold: a non-negligible fraction of these impurities occupy interstitial positions at equilibrium, and the activation for the jumping of these interstitial atoms quite low. One of the first studies of this rapid diffusion dealt with the low temperature diffusion of Cu in Ge, and the fact that the diffusion did not appear to obey Fick's Laws.

Before describing these experiments it is necessary to emphasize a fundamental difference between metal samples and the crystals of Ge or Si used in such studies. The density of dislocations, may average $1/cm^2$ or less in a good Ge or Si crystal, while in a metal the density is high ($10^5/cm^2$) even in a well annealed crystal. Thus in a metal the distance a vacancy must diffuse to find a dislocation is much shorter than in a semiconductor. As a result, in a metal the equilibrium concentration of vacancies is usually maintained throughout the crystal. In Ge or Si, vacancy sources are far apart, and the concentration of vacancies or interstitials may deviate from the equilibrium value over a large fraction of the crystal.

Now return to the Cu-Ge experiment in which diffusion does not appear to obey Fick's Law. If Cu is deposited on one face of a Cu free Ge crystal containing only a few dislocations in one sub-boundary, a diffusion anneal leads to the copper distribution shown in Fig. 5-12. There is a high concentration on the surface, then a low copper concentration where only interstitial Cu is found in the lattice, then a rise in Cu concentration along the sub-boundary. Interstitial Cu is converted into a substitutional atom in the Ge lattice by the reaction

$$Cu_{Ge} = Cu_I + V_{Ge} \quad K = N_i N_v / N_s, \quad (5\text{-}40)$$

This is termed the 'dissociative mechanism.' At temperatures where diffusion is easily measured, 700-800° C, the equilibrium constant K for this reaction is much less than one. This means that the solubility on substitutional sites, N_s^e, is much greater than that on interstitial sites, N_i^e. However, the low concentration of interstitial Cu that develops near the Cu rich surface rapidly spreads through the perfect crystal since the diffusion coefficient of the Cu interstitial D_i is quite large.

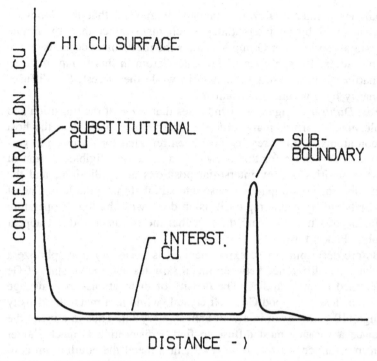

Fig. 5-12—$C(x)$ for Cu diffused into a Ge crystal containing a sub-boundary.

(More precisely $D_i N_i \gg D_v N_v$). In regions far from vacancy sources the concentration of copper can only rise to N_i^e since the concentration of vacancies N_v is very small ($N_i^e \gg N_v$) and there is no source of additional vacancies. However, near dislocations, and free surfaces, where new vacancies can be produced and N_v maintained at thermal equilibrium, the conversion of interstitial Cu to lattice sites will continue until the Cu concentration on lattice sites equals that on the outside of the crystal. In the crystal shown in Fig. 5-12 the concentration at the sub-boundary has not yet risen this high.

It is useful to treat the diffusion of these fast diffusing elements by considering two limiting cases, that for perfect crystals where the surface is the only source of vacancies, or sink for interstitials, and that for crystals with many dislocations in which local equilibrium is maintained throughout the crystal. In each of these cases the conversion of solute atoms from interstitial to lattice sites can be accomplished by two mechanisms, the kick-out mechanism given in Eq. (5-39) or the dissociation mechanism, Eq. (5-40).

A detailed analysis of the effect of these conversion mechanisms on the effective diffusion coefficient of the solute D_{eff}, and on $c(x,t)$, allows one to distinguish between the two conversion mechanisms.[17]

If the crystal is dislocation free, the vacancies needed for the *dissociation reaction* in Eq. (5-40) can only move in from the free surface. Consider a crystal with pure copper on the surface. After a brief diffusion anneal the perfect crystal may be permeated with interstitial copper, but the equilibrium concentration of copper, N_s^e, only develops as fast as vacancies can diffuse in from the surface to convert interstitial Cu to substitutional atoms. Under these conditions the advance of the copper 'front' is determined by a balance between the flux of vacancies from the surface, $D_v N_v^e$, and the amount of copper needed to advance the front unit distance, that is the solubility $N_s^e - N_i^e \simeq N_s^e$. The effective diffusion coefficient of Cu is thus given by

$$D_{eff} = D_v N_v^e / N_s^e \quad \text{(disloc. free)} \quad (5\text{-}41)$$

where the superscript 'e' indicates the value at thermal equilibrium. From the known solubility of Cu in Ge, N_s^e, the value of $f_v D_v N_v^e$ can be calculated. This is found to equal the tracer self diffusion coefficient of Ge, as it should.

At the other extreme, if the crystal has a high density of dislocations then the vacancy concentration is maintained at its equilibrium value N_v^e throughout the crystal. The flow of copper in from the copper bearing surface is determined by the flux of interstitials $D_i N_i^e$. The amount of copper needed to advance the front a distance dx is the solubility of substitutional copper at thermal equilibrium. Thus D_{eff} for Cu is

$$D_{eff} = D_i N_i^e / N_s^e \quad \text{(dislocations)} \quad (5\text{-}42)$$

D_{eff} in this case is much greater than for the dislocation free crystal, because $N_i^e \gg N_v^e$. Eq. (5-42) is similar to equation (3-34) for the diffusion of hydrogen through hydrogen free iron. That is, mobile interstitial Cu corresponds to the hydrogen diffusing rapidly through the lattice, and the Cu on substitutional sites corresponds to the hydrogen in traps.

The *kick-out mechanism* is the other mechanism to move interstitial solute onto lattice sites. In this case the impurity interstitial has a higher energy than that of the Cu discussed above, and instead of waiting for a vacancy, it can move an Si atom off of its lattice site and move it into an interstitial position. (It is believed that Au diffusing in Si proceeds in this way.) If the kick-out mechanism operates, Eq. (5-39), a lot of interstitial Si atoms are produced and the process limiting the influx of solute may be the elimination of these slowly diffusing silicon

---

[17]W. Frank, et al., loc. cit., pp. 116–35.

interstitials. In a crystals containing many dislocations the removal of excess Si interstitials is rapid (N_I^e is maintained). Solute interstitial diffusion is then limiting and Eq. (5-42) again gives the effective solute diffusion coefficient.

The case of solute diffusion in dislocation free crystals is more interesting because here the kick-out mechanism gives results different from the dissociative mechanism. Again interstitial solute permeates the crystal, and now the Si interstitials generated by the kick-out reaction must diffuse to the free surface before more interstitial solute move into the region. In analogy with Eq. (5-41), one would expect an equation of the form,

$$D_{eff} = D_I N_I / N_s. \qquad (5\text{-}43)$$

where N_I is the atom fraction of Si interstitials. However, for the case of solute diffusion into a pure Si crystal, the ratio N_I/N_s rises substantially from a low value at the free surface to a high value in the interior. Thus D_{eff} increases substantially in moving from the free surface into the front where the concentration is changing. This rise produces a flat concentration gradient and leads to the unique 'signature' of this mechanism. This can be seen as follows. Remember that the rapidly diffusing solute interstitial has an essentially constant concentration N_i throughout the couple. The equation for the equilibrium constant K in Eq. (5-39) can then be rewritten

$$KN_i = N_s^e N_I^e = N_s N_I \qquad (5\text{-}44)$$

Substitution for N_I in Eq. (5-43) gives

$$D_{eff} = (D_I N_I^e N_s^e)/(N_s)^2 \qquad (5\text{-}45)$$

At the free surface the value of D_{eff} is $D_I N_I^e / N_s^e$. However, well beneath the free surface, the conversion of interstitial solute atoms generates a concentration of Si interstitials well in excess of the equilibrium value ($N_I \gg N_I^e$), as N_s falls. As a result D_{eff} rises to a much higher value. The diffusion equations have been solved for this variable D_{eff}, and the predicted slow drop off in the observed $c(x,t)$ curve fits the observed data for Au diffusing in Si quite satisfactorily.[18]

5-9 ORDERED ALLOYS AND INTERMETALLIC PHASES[19]

Ordered alloys and intermetallic compounds form the bridge between dilute alloys in which the tendency toward short-range order is

[18]W. Frank, et al., op. cit., pp. 125–8.
[19]H. Bakker, *Diffusion in Crystalline Solids*, ed. G. E. Murch, A. S. Nowick, Academic Press (1984), p. 189–258.

weak, and the ionic compounds in which the energy to put an A atom on a B site in the compound AB is so large that each type of ion diffuses only on its own sublattice. Such alloys are always ordered at low temperature, and sometimes this order persists up to the melting point. In an ordered alloy the energy of an A atom on a B-type site (β site) is higher than that of an A atom on an A-type site (α site). However, this energy difference is small enough that some A atoms do reside on β sites even in the ordered phase. This can be seen from the existence of the ordered phase over a range of compositions or from the lack of perfect long-range order in stoichiometric intermetallic phases. The study of diffusion in such phases involves two difficult problems. First, the vacancy concentration will depend on the composition of the phase, so to understand the diffusion mechanism one must determine to what extent a deviation from stoichiometry toward an excess of A is accommodated by placing A atoms on β sites and to what extent by forming vacant β sites. Second, if A atoms diffuse in the ordered phase by moving into vacant adjacent β sites, there will be a high probability of the A atom returning to the vacant α site on its next jump. This correlation effect can be quite strong. The main question to be discussed here is the mechanism of diffusion in stoichiometric ordered alloys. Questions such as defect types in off stoichiometric compositions,[20] or the interesting matters of creep and dislocation climb in such phases, are not addressed.

Consider the ordered intermetallic phase shown in Fig. 5-13. This is often referred to as the CsCl or ordered beta brass structure. In it A atoms have only B nearest neighbors, and B atoms having only A

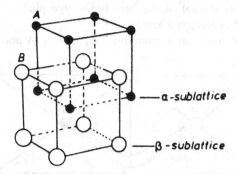

Fig. 5-13—Two cells of the CsCl structure. If the alloy is disordered, the lattice is bcc. (From H. Bakker)

[20]J. P. Neumann, *Acta Met.*, *28* (1980) 1165–70. H. Bakker, et al. in *Atomic Transport in Alloys: Recent Development*, Ed. G. E. Mirch, M. A. Dayananda, TMS/AIME (1985), pp. 39–66.

nearest neighbors. When the alloy is disordered the phase has a bcc structure. If diffusion occurred by jumps to vacancies on nearest neighbor sites then diffusion would destroy the order of the crystal. Three mechanisms that would allow diffusion and preserve the order are:[21]

(a) an atomic jump to a vacancy on a next nearest neighbor site, that is, an A atom jump to a vacant a site along a $\langle 110 \rangle$ direction.

(b) a pair of vacancies, one on each sublattice, start on adjoining sites. (See Fig. 5-14) One vacancy makes a $1/2\langle 111 \rangle$ jump which dissociates the pair. This places an A atom on a β site, or a B on an α site. If the misplaced atom now jumps into the second vacancy it returns to its proper sublattice and the divacancy is reassociated. Note that the vacancy pair can give diffusion on either sublattice.

(c) one vacancy, on either sublattice, can make a sequence of six correlated nearest neighbor jumps in a ring in a $\{110\}$ plane so as to exchange the two A atoms on their sublattice and the two B atoms on their sublattice, Fig. 5-15. Though the sequence begins and ends with perfect order, after only three jumps there are three misplaced atoms. This sequence is the simplest (introduces the least disorder) while allowing diffusion with only one vacancy and nearest neighbor jumps.

The ratios of D_A/D_B for each of these mechanisms differs. For (a) and (b) the ratio can have any value since the diffusion on the two sublattices is uncoupled, while for (c) the ring cannot operate without a coupling between jumps on the two sublattices. A detailed analysis shows the ratio must lie between 0.5 and 2.0 for the six-jump-ring. A ratio within this range extending over a wide temperature range has been observed in several of the beta brass type phases, e.g. CoGa, AuZn, AgMg, to mention a few.

The energy increase on exchanging neighboring A and B atoms in

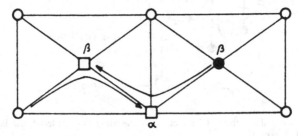

Fig. 5-14—Diffusion by a divacancy mechanism shown on the (110) plane of a CsCl lattice. β indicates the B-type sites in the ordered alloy. (Hahn, Frohberg, & Wever)

[21]H. Hahn, G. Frohberg, H. Wever, *Phys. Stat. Sol.(a)*, 79 (1983) 559–65.

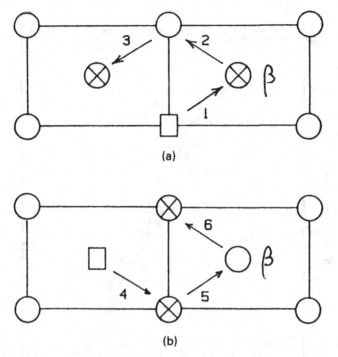

(a)

(b)

Fig. 5-15 — (a) shows the initial configuration of a six-jump-ring mechanism on a {110} plane and arrows representing the 3 vacancy jumps that lead to (b). The arrows in (b) show the jumps that will complete the ring. ⊗ are B atoms, ○ are A atoms.

an initially ordered alloy is called the disordering energy, V_d. If this disordering energy is less than the mean thermal energy of the atoms at the melting point, i.e. V_d/RT_m is less than one, the phase will disorder at some temperature below the melting point of the alloy. If it is high the intermetallic phase will remain ordered to its melting point. The correlation factor f for diffusion in an ordered alloy decreases rapidly as V_d/RT increases, that is as the temperature drops. This can be seen in Fig. 5-16 where diffusion data are shown for beta brass in the temperature range that includes the transition temperature. Much of the increase in the activation energy for diffusion as the temperature drops below the ordering temperature is due to a rapid decrease in f with rising V_d/RT. In systems where the disordering energy is quite high, f for nearest neighbor jumps goes so low that nearest neighbor jumps stop contributing to diffusion. Then the second nearest neighbor jump mechanism probably predominates. Cation diffusion NaCl or CoO would be extreme examples of this situation.

The other type of ordered intermetallic that has been studied exten-

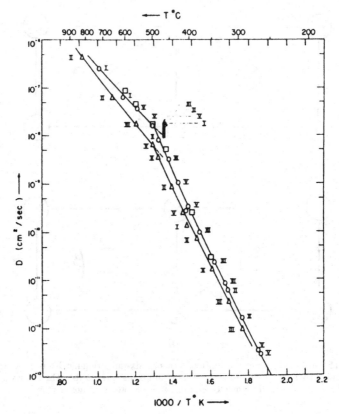

Fig. 5-16—D for Cu(Δ) and Zn($\circ$) in β-CuZn vs. $1/T$ in alloys ranging from 46.5 to 48 a/o Zn. [A. Kuper, et al., *Phys. Rev.*, *104* (1956) 1536.]

sively is of the type Ni$_3$Sb. Here the Ni type atoms occupy two kinds of sites as shown in Fig. 5-17, but the majority component (Ni in this case) can diffuse without disturbing the order of the alloy. The majority component is found to diffuse with a low activation energy relative to the melting point of the compound. Also, it moves with an appreciably lower Q than the minority element. For example in Cu$_3$Sn, $Q_{Cu} = 0.85$ eV and $Q_{Sn} = 1.11$ eV in a 79.8 a/o Cu alloy.[22] In Cu$_3$Sb, which melts at only about 10% lower temperature, $Q_{Cu} = 0.31$ eV. At least part of the explanation for this is the tendency to form structural vacancies on the majority component sublattice with deviations from stoichiometry.

---

[22]N. Prinz, H. Wever, *Phys. Stat. Solidi A*, *61* (1980) 505, see also Bakker.

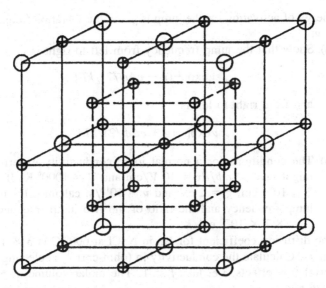

Fig. 5-17—The A_3B structure. Larger circles correspond to B atoms. Smaller to A atoms. The lattice is bcc when disordered.

PROBLEMS

5-1. The free energy of a diffusing ion of charge e decreases by $\alpha e(d\phi/dx) = e\Delta\phi$ in moving from left to right in the figure. If the free

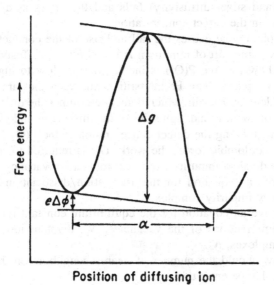

Position of diffusing ion

energy of activation for the jump is g and the vibration frequency is v:

(a) Show that the jump frequency from left to right is

$$v \exp[-(g - e\Delta\phi/2)/kT]$$

and from right to left is

$$v \exp[-(g + e\Delta\phi/2)/kT].$$

(b) The conditions for a normal ionic conductivity experiment might involve $d\phi/dx = 10$ V/cm and $T = 1000°$ K. If $\alpha = 2 \times 10^{-8}$ cm, $g = 2$ eV and $v = 10^{14}/s$, calculate the mean jump frequency and the ratio of the two jump frequencies. ($k = 8.7 \times 10^{-5}$ eV/K

5-2. The diffusion coefficient for Na in NaCl at $600°$ C is 3×10^{-10} cm^2/s. Calculate the conductivity in (ohm-cm)$^{-1}$, neglecting any correlation effects, i.e. take $f = 1$. The molar volume of NaCl is 26 cm^3

5-3. (a) Define Frenkel and Schottky disorder.
   (b) Give the equations dictated by charge neutrality relating the defect concentrations for each case.
   (c) Find the ratio N_{vc}/N_{va} for the case where the free energy to form a Schottky pair (G_S) equals that to form a Frenkel pair (G_F).

5-4. As Cd is dissolved in initially pure AgBr the electrical conductivity (and silver diffusivity) falls and then rises as a result of changes in the defect concentration.
   (a) Explain what causes the fall and rise of the conductivity.
   (b) The same sort of minimum is found for the diffusivity of Fe in Fe_3O_4 as the $P(O_2)$ is increased from low to high in the Fe_3O_4 phase field. Fe interstitials and vacancies are the type of defects. Explain which defect determines the diffusivity of Fe on the low and high sides of the minimum, and give equations relating the defect concentration to the $P(O_2)$.

5-5. Due to coulombic forces the work, G_p, is required to reversibly move a divalent impurity from next to a cation vacancy in NaCl.
   (a) Write an equation for the association/dissociation of a vacancy-impurity complex.
   (b) Derive an equation for the equilibrium constant relating the atom fractions of the vacancies, N_p, divalent ion, N_d, and complexes, N_c.
   (c) How should the number of nearest neighbors on the cation sub-lattice enter this equation.

5-6. The most pronounced differences between the tracer diffusion coefficients of the more mobile component, D_A^* and the chemical $\tilde{D}$ are found in compounds. Here the difference is due primarily to the rapid change of activity with stoichiometry in crossing a phase field. Thus one customarily finds that $\tilde{D} \gg D_A^* \gg D_B^*$ where each "$\gg$" reflects changes of several orders of magnitude.[23]

Consider a binary system of the elements A and B with one compound AB. Assume the terminal solid solutions (A in B, and B in A) to have virtually no solid solubility and be ideal ($g_i = 1$ in $a_i = g_i N_i$).

(a) Graph the activity of A, a_A, vs. N_A for $N_A = 0$ to $N_A = 1$.

(b) Estimate the average magnitude of $d \ln(a_A)/d \ln(N_A)$ in the AB compound if the stoichiometry range of the compound is 10^{-5}.

(c) The variation of a_i across the AB phase field is not linear but sigmoidal with the most rapid change at exact stoichiometry. If the type of disorder is by the formation of Frenkel pairs, would this slope increase or decrease with increasing G_F?

5-7. Explain the relationship between the following for a compound semiconductor like Fe_3O_4

(a) Cation vacancy concentration and oxygen partial pressure.

(b) Cation interstitial concentration and oxygen partial pressure.

(c) Diffusion coefficient of the cation and the oxygen partial pressure.

5-8. In a commercial silicon single crystal there are virtually no dislocations present. For fast diffusing impurities (dopants) this often gives rise to a lattice diffusion coefficient which varies with the dislocation content of the crystal. Explain how and why this is true.

Answers to Selected Problems

5-1. (b) jump freq. $= 10^4$/s, ratio $= \exp[e\Delta\phi/kT] = 1 + 2 \times 10^{-7}$

5-2. Use Eq. (5-10) $D_\sigma = \sigma RT\Omega/F^2$, $\sigma = 1.5 \times 10^{-5}$ (ohm-cm)$^{-1}$

5-3. (c) From charge neutrality, $N_{ic} + N_{va} = N_{vc}$, but $N_{ic} = N_{vc}$ so $N_{vc}/N_{va} = 2$.

5-4. (a) See Sec. 5.5. (b) See Sec. 5.7.

---

[23]See for example K. Becker, H. Schmalzried, V. Wurmb, *Solid State Ionics, 11* (1983) 213–9.

6

HIGH DIFFUSIVITY PATHS

In the preceding chapters the only defects which aided diffusion through the crystal were vacancies and interstitials. Dislocations, free surfaces, and grain boundaries entered only to help attain the equilibrium defect concentration. However, it is now well established that the mean jump frequency of atoms at dislocations, boundaries, or surfaces is much higher than that of the same atom in the lattice. The diffusivity is therefore higher in these regions. This higher diffusivity is of interest for several reasons. First, there is the question of what error these paths introduce in the measurement of the lattice diffusion coefficient. Also, with properly designed experiments it is possible to determine the diffusion coefficients in each of these high diffusivity regions, allowing one to learn more about the structure of these paths and about how the atoms move in them. Finally, there are a group of kinetic processes which are limited by such diffusion, for example diffusional creep, structural changes in thin films, or the stability of fine catalysts.

As an example of the phenomena we are talking about, the contribution of diffusion along grain boundaries can be seen in Fig. 6-1. Here the apparent self-diffusion coefficient in silver is shown for single-crystal and polycrystal samples. This apparent diffusion coefficient is just that value of D obtained by plating radioactive silver on the surface of the specimen, diffusing it, and then determining D from a plot of ln(activity) vs. penetration distance squared. At high temperatures, the same value of D is obtained from both types of samples. However, below 700° C the values of D obtained using a polycrystal consistently lie above the values obtained with a single crystal. The high-diffusivity paths[1] in this case are grain boundaries. Below about

---

[1]Zener coined the phrase "short-circuiting paths" to describe this type of effect. For one accustomed to thinking in terms of the electrical analogue of diffusion, it is apparent that a high-diffusivity path corresponds to one of high conductivity, and this will tend to relieve the potential gradient and act like a short circuit. However, this analogy is not immediately apparent to many, and the phrase "high-diffusivity path" is used here.

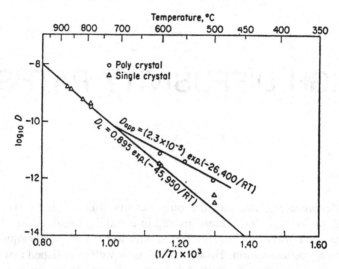

Fig. 6-1 — Values of D_T for silver in silver single-crystals and polycrystalline samples. [D. Turnbull, in *Atom Movements*, ASM, Cleveland, (1951) p.129.]

600° C ($0.7T_m$) the contribution of the grain boundary region becomes dominant.

An estimate of the increase in the jump frequency in the neighborhood of a grain boundary can be obtained as follows. In pure silver the smallest grain diameter which can be retained at high temperatures will be about 1 mm. If this is true and the high-diffusivity region around a grain boundary is taken to be 3×10^{-8} cm wide, about one atom in 10^6 will be in the grain boundary. At 650° C these few grain boundaries double the measured diffusion coefficient. If one-millionth of the atoms make a contribution to the jump frequency which is comparable to that of all the rest of the atoms, then each of these must be jumping roughly one million times as often as the regular lattice atoms. At lower temperatures the difference between the jump frequency in the grain boundaries and that in the lattice is even larger. Since the grain boundary atoms represent such a small part of the specimen, it also follows that the mean jump frequency in this region can be a few orders of magnitude larger than it is in the lattice, for example 10^3 times, and still the boundary regions will make no significant contribution to the total flux.

The first problem to be dealt with is how to measure the diffusion coefficient in these high-diffusivity paths. These paths cannot exist except as regions surrounded by otherwise perfect crystals, so some means must be found of treating measurements made on samples in which the surface atoms represent a very small fraction of the atoms in the sample. Two types of such experiments are discussed. The first uses a concentration gradient as a driving force and the accumulation of solute to measure the total amount of material transported. The second uses surface tension as a driving force and obtains the total flux from the change in the shape of the sample.

6.1 ANALYSIS OF GRAIN BOUNDARY DIFFUSION

To obtain values of the grain boundary diffusion coefficient D_b from diffusion studies on bicrystals, Fisher suggested the following analysis.[2] Consider the grain boundary to be a thin layer of high-diffusivity material between two grains which have a low diffusivity. A section normal to the grain boundary and the free surface is shown in Fig. 6-2.

To obtain the differential equation which is valid inside the high diffusivity slab, consider an element of this slab which is dy long by δ thick by unit length deep (into the page in Fig. 6-2). The fluxes into, or out of, the faces normal to the x and y axes are shown in Fig. 6-3. Any plane normal to the z axis is a symmetry plane, so J_z would equal zero. A dimensional argument shows that the rate of change in concentration for this element of grain boundary is given by

---

[2]J. C. Fisher, *J. Appl. Phys.*, **22** (1951) 74.

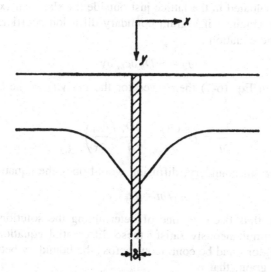

Fig. 6-2—Coordinate axes and isoconcentration line in a section of the model used for grain boundary diffusion analysis.

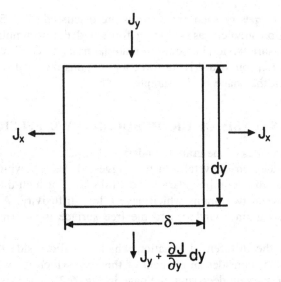

Fig. 6-3—Fluxes into and out of an element of grain boundary slab.

$$\frac{\partial c_b}{\partial t} = \frac{1}{1 \, dy \, \delta} \left[\delta \left(J_y - J_y - \frac{\partial J_y}{\partial y} dy \right) - 2 \, dy \, J_x \right] = \frac{-\partial J_y}{\partial y} - \frac{2}{\delta} J_x \quad (6\text{-}1)$$

J_x is the flux out of the grain boundary into the perfect lattice and can be replaced by $-D_1(\partial c_1/\partial x)$ where D_1 is the lattice D, and the gradient is evaluated in the lattice just outside the slab. An expression for J_y can be obtained if a grain boundary diffusion coefficient D_b is defined by the equation

$$J_y = -D_b \, \partial c_b/\partial y \qquad\qquad (6\text{-}2)$$

Substituting in Eq. (6-1) then gives for the $c(x,y,t)$ in the boundary slab

$$\frac{\partial c_b}{\partial t} = D_b \frac{\partial^2 c_b}{\partial y^2} + \frac{2D_1}{\delta} \left(\frac{\partial c_1}{\partial x} \right)_{x=\delta/2} \qquad (6\text{-}3)$$

Outside the grain boundary, diffusion would obey the equation

$$\partial c_1/\partial t = D_1 \nabla^2 c_1 \qquad\qquad (6\text{-}4)$$

The problem thus becomes one of determining the solution $c(x,y,t)$ which will simultaneously satisfy these differential equations in the respective regions and be continuous across the boundary between the slab and the grain, that is

$$c_b(\delta/2) = c_1(\delta/2) \qquad\qquad (6\text{-}5)$$

Solutions. Experimentally, solute is applied to the free surface at y = 0 and allowed to diffuse into the sample. Solutions have been obtained for the boundary conditions of either constant surface composition, or the application of a thin film to the surface. If D_b is much greater than D_1 the solute penetration around the grain boundaries will be deeper than through the lattice. It is the distribution of this material which diffused along the boundary to a depth well beneath the layer entering through the lattice that allows the determination of D_b. Working with the case of constant surface composition, Fisher found that the concentration in the grain boundary rises quickly at first but then at an ever-decreasing rate. Thus the grain boundary concentration at any point on the boundary will be near its final value during much of the anneal. To simplify the analysis he assumed that the boundary composition at each point stays at its final value throughout the experiment and that the flux in the lattice is perpendicular to the boundary. This yields a simple, approximate solution for the solute that enters the lattice via the boundary. The same type of assumptions have been used by others in considering other geometries.

The first exact solution, and the one most frequently used, is due to Whipple. He also assumed a constant surface composition, and used a Fourier-Laplace transform to obtain a solution in integral form.[3] Numerical analysis indicated[4] that beneath the surface layer due to lattice diffusion where $dln(\bar{c})/d(y^2)$ is constant, Whipple's solution gives a region where the slope $dln(\bar{c})/d(y^{6/5})$ is constant and $D_b\delta$ can be obtained from the equation[5]

$$[dln(\bar{c})/d(y^{6/5})]^{5/3} = 0.66(D_1/t)^{1/2} (1/\delta D_b) \qquad (6\text{-}6)$$

Note that one can determine only the product $D_b\delta$, not D_b alone. This equation is valid only in the region of y greater than, say, $4\sqrt{D_1 t}$, where the tracer present has entered the crystal by diffusing in along the boundary and then out into the crystal from grain boundary. If such a region is to develop, D_b must be much greater than D_1. Whether or not it develops is indicated by the parameter

$$\beta = (D_b/D_1)(\delta/2\sqrt{D_1 t}) \qquad (6\text{-}7)$$

which should be 10 or greater for Eq. (6-6) to be used. Figure 6-4 is adapted from Whipple and shows contours for a concentration 0.2 times the surface concentration, for $\beta = 0.1$, 1.0, and 10. Note that there is no significant extra penetration along the grain boundary until $\beta >$

[3]R. T. Whipple, *Phil. Mag.*, *45* (1954) 1225.
[4]H. Levine, C. MacCallum, *J. Appl. Phys.*, *31* (1960) 595.
[5]See N. L. Peterson, in *Grain Boundary Structure and Kinetics*, ASM, Metals Park, OH (1980), pp. 209–37.

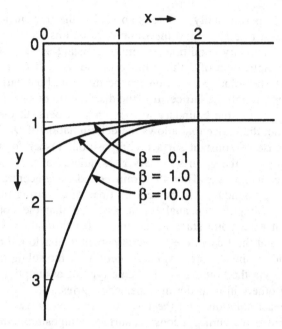

Fig. 6-4—Iso-concentration contours showing the degree of extra penetration near grain boundary for various values of β. $y = 1$ is the penetration in the absence of a grain boundary. (R. T. Whipple)

1. If $\delta = 4 \times 10^{-8}$ cm, and $D_1 = 10^{-11}$ cm$^2/s$, then $t = 10^5$ s (≈ 28 hr). This means that D_b/D_1 must be greater than 5×10^4 before there is appreciable penetration at the boundaries. Physically the reason for this is that the grain boundary slab is so thin that the grain boundary flux is not sufficient to bring in enough material to distort the contours until $D_b/D_1 > 5 \times 10^4$.

Measurement of D_b. If D_b is much greater than D_1, solute will diffuse along the boundary to a greater depth than in the lattice before being drained off into the grains by lattice diffusion. Two methods for determining D_b have been used extensively:

- measure the distribution of in-diffused solute in a series of thin slices cut parallel to the sample surface, $c(y)$, and use Eq. (6-6) to give $D_b\delta$.
- measure the depth of penetration of a given concentration at the boundary Δy compared to the lattice penetration from the surface far from the boundary.

The second technique is useful for comparing the relative $D_b\delta$ of different boundaries. The first is used for accurate determination of $D_b\delta$.

Experimentally one invariably applies a thin film of tracer to the surface rather than maintaining a constant surface concentration as assumed in Whipple's solution. Suzuoka found the solution for the thin film case.[6] The solution differs from that of Whipple, but the slope of $dln\bar{c}/dy^{6/5}$ is nearly the same. If $\beta \geq 10$ the equation for determining $D_b\delta$ given by Suzuoka's solution results only in the constant 0.66 in Eq. (6-6) being decreased by a few percent.

A thin-film tracer experiment in a crystal free of grain boundary effects produces a penetration curve with $ln(\bar{c})$ vs y^2 being a straight line. If grain boundary diffusion makes a significant contribution, the tracer penetrates much deeper, the penetration curve drops more slowly and gives a line for $ln(\bar{c})$ vs $y^{6/5}$. Thus the penetration curve consists of a sum of two terms and can be described by the equation

$$c(y,t) = A_I \exp(-y^2/4Dt) + A_{II} \exp(-y^{6/5}/b) \qquad (6\text{-}8)$$

where A_I, A_{II}, and b are known constants. If $ln(\bar{c})$ drops more slowly than y^2, this is a clear indication that some high diffusivity path is operating in addition to lattice diffusion. Fig. 6-5 shows a penetration curve for a gold tracer diffused into a fine grained thin film of gold

---

[6]T. Suzuoka, *J. Phys. Soc. Japan, 19* (1964) 839.

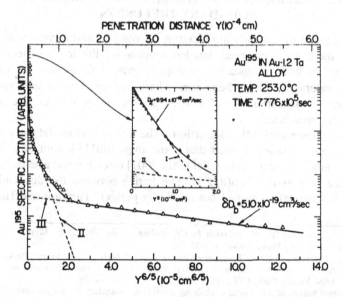

Fig. 6-5—Penetration plot for Au* into polycrystalline Au. A three term exponential fitting procedure yields three lines, and values of D for diffusion in the lattice (inset, I), sub-grains (II) and grain boundaries (III). [D. Gupta, *J. Appl. Phys., 44* (1973) 4455.]

at $253°$ C $(0.4\ T_m)$. The data has been fit to an equation of the form of Eq. (6-8) but with an additional $y^{6/5}$ term. The small inset shows that the data points closest to the surface fit $ln(\bar{c})$ vs y^2. The main graph is a plot of $ln(\bar{c})$ vs. $y^{6/5}$ with two straight lines, labeled II and III. The one main line (III) is due to diffusion along grain boundaries, and clearly penetrates quite deeply.

Eqs. (6-5) and (6-6) are only valid for bicrystals, or polycrystal-line samples for which the grain diameter $2R$ is much greater than the mean diffusion distance in the lattice $2\sqrt{D_l t}$. If R is less than or equal to $\sqrt{D_l t}$, then different effects are seen. These are discussed in Sec. 6-3 along with the effect of arrays of dislocations.

Before leaving this analysis it should be pointed out that the mathematical analysis given above is also applicable to the case of surface diffusion. The slab in Fig. 6-2 is a plane of symmetry so there will be no net flux across it. Thus if the half of the bicrystal to the left of the slab is removed, the high-diffusivity slab remains, but now it corresponds to a solid-vapor interface. In the derivation of Eq. (6-3), the only change required is to remove the factor of 2 since the volume element in Fig. 6-3 now loses material to the lattice on only one side.

6.2 EXPERIMENTAL OBSERVATIONS ON GRAIN BOUNDARY DIFFUSION

As a result of transmission electron microscopy and computer modeling, there are now quite detailed models for the low temperature structure of grain boundaries, and quite reasonable models for their high temperature kinetic behavior.[7] There has also been a substantial number of measurements of $D_b\delta$ as a function of temperature and boundary structure.[8]

Grain Boundary Misorientation. The dislocation model for a low angle grain boundary predicts that a low angle [001] tilt boundary consists of edge dislocations parallel to the [001] direction and a separation distance of $s = a_o/[2\sin(\theta/2)]$.[9] The lattice between the dislocation cores is elastically strained but relatively perfect. Turnbull and Hoff-

---

[7]See R. W. Balluffi, in *Diffusion in Crystalline Solids*, ed. G. E. Murch, A. S. Nowick, Academic Press, 1984, p. 320–78.

[8]N. L. Peterson, in *Grain Boundary Structure and Kinetics*, ASM Metals Park, OH (1980) 209–37. And, G. Martin, B. Perraillon, in *Grain Boundary Structure and Kinetics*, ASM Metals Park, OH (1980) 239–95.

[9]The two halves of a bicrystal containing a (001) tilt boundary have a common [100] direction in the plane of the boundary and can be brought into coincidence by rotation about the [100]. If the grain boundary also is a plane of symmetry, it is said to be a symmetric tilt boundary. If the common [100] direction is normal to the boundary, the boundary is called a (100) twist boundary.

man postulated that in the core of these dislocations the diffusion coefficient D_p is much greater than D_l. Thus instead of replacing the grain boundary by a slab of uniform thickness δ and diffusivity D_b, the boundary is described as a planar array of 'pipes' of radius 'a' and spacing s. For diffusion in the direction of the dislocation cores or pipes this gives the equation

$$p = D_b\delta = D_p(\pi a^2/s) = D_p\pi a^2[2\sin(\theta/2)/a_o] \simeq D_p\pi a^2\theta/a_o \quad (6\text{-}9)$$

Several predictions stem from this model.

- $D_b\delta$ should increase linearly with misorientation, at least in the low angle region where the dislocation cores don't overlap. This is borne out by the data in Fig. 6-6.
- The activation energy for grain boundary diffusion $Q_b = Q_p$ should be independent of θ, at least in the low angle region. (See Fig. 6-6.)
- $D_b\delta$ should not be isotropic in the boundary, but should be appreciably larger in the direction of the pipes than normal to them. As θ increases, and the cores get closer together, the anisotropy should decrease. The data in Fig. 6-7 shows this. Also, the anisotropy persists even at very high angles where the dislocation model is no longer valid. Similar results have been demonstrated for other metals.

The data for silver in Fig. 6-6 and Eq. (6-9) lead to the equation

$$D_p = 0.1 \exp(-82,500/RT) \text{ cm}^2/s \quad (6\text{-}10)$$

for the pipe diffusivity. This is consistent with the data for high angle

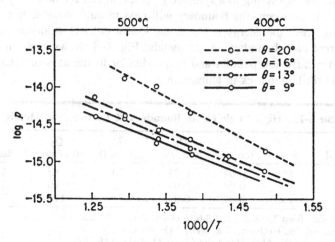

Fig. 6-6 — Dependence of $\log(D_b\delta)$ on temperature and θ for [100] tilt boundaries in Ag. [D. Turnbull, R. Hoffman, *Acta Met.*, 2 (1954) 419.]

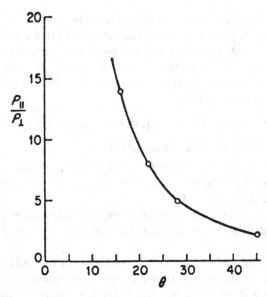

Fig. 6-7—Anisotropy of $D_b\delta$ on θ for [100] tilt boundaries in Ag. [R. Hoffman, *Acta Met.*, *4* (1956) 98.]

boundaries in Table 6-1. Using Eq. (6-10) and the data in Table 6-1 the ratio D_p/D_1 at $T/T_m = 0.7$, 0.5, and 0.3 are, 5×10^5, 6×10^7, and 6×10^{12}, respectively. These are typical of other metals.

At larger misorientations the dislocation model is no longer valid, and as θ keeps increasing one ultimately rotates the two crystals through an angle corresponding to a symmetry operation and the boundary disappears. In between, the boundary will pass through coincidence orientations where the energy is lower, the lowest order coincidence being a coherent twin boundary in fcc metals. Fig. 6-8 shows such an example for diffusion parallel and perpendicular to the axis of rotation in [011] tilt boundaries in aluminum.

Table 6-1.   High Angle Grain Boundary & Lattice D in Metals

Metal	Struc	$D_{\varrho l}$ (cm^2/s)	$D_{\varrho b}$ (cm^2/s)	Q_l (kJ/mol)	Q_b (kJ/mol)	Ref.
Ag	fcc	0.04	0.03	170	85	1
Au	fcc	0.04	0.03	170	90	2
Ni	fcc	0.92	0.07	278	115	3

Lattice data from Table 2-5 and 2-6. δ assumed 3×10^{-8} cm.
[1]D. Turnbull, R. Hoffman, *Acta Met.*, 2 (1954) 419.
[2]D. Gupta, K. W. Asai, *Thin Solid Films*, 22 (1974) 121.
[3]A. R. Wazan, *J. Appl. Phys.*, 36 (1965) 3596.

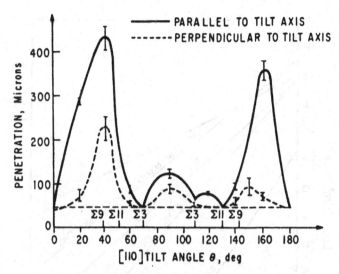

Fig. 6-8—Penetration of Zn parallel and perpendicular to the [011] rotation axis of Al bicrystals. [I. Herbeuval, M. Biscondi, C. Goux, *Mem. Sci. Rev. Met.*, *70* (1973) 39.]

Clearly there is a variation with angle and a minimum at the coherent twin orientation ($\Sigma = 3$). The lack of symmetry reflects scatter in the data and scatter in the misorientation of the bicrystals used.

Temperature Dependence. Experimental results clearly indicate that the activation energy for grain boundary diffusion is less than that for lattice diffusion, while D_o is about the same for the two. This is born out by the data in Table 6-1. The samples used to get this data were polycrystalline and the observed penetration will be weighted toward high angle grain boundaries with higher values of $D_b\delta$.

If one compares $D_b\delta$ for different types of grain boundaries, the results can be rationalized in terms of the atomic packing in the given boundary compared to the packing (density) in a perfect lattice. If there is more open space a grain boundary, some of the atoms will be able to jump with a lower activation energy than if the atoms are more tightly packed. A lower Q will translate into a higher value of D. Such behavior is shown in Fig. 6-9 which represents the spectrum of data for D in: boundaries made of undissociated dislocations or high angle boundaries ($Q_b/Q_1 = 0.4–0.5$), dissociated dislocations and twist boundaries ($Q_b/Q_1 = 0.6–0.8$), and lattice diffusion. For the same misorientation between two grains, diffusion down a twist boundary is 10 to 100 times slower than down the tilt boundary. Examination of a tilt boundary shows that it has clear regions of low density while a twist boundary involving screw dislocations has little dilatation and primary shear strains. (The surface diffusion curve is discussed below.)

Diffusion in Solids

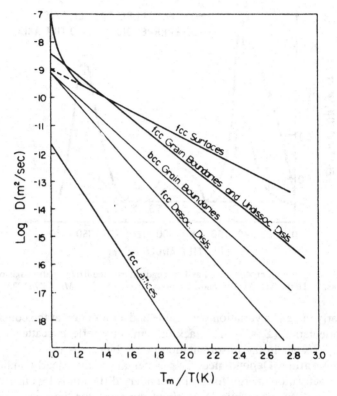

Fig. 6-9 — Diffusivity vs. reduced reciprocal temperature for various diffusion paths in metals. [N. Gjostein, in *Diffusion*, ASM, Metals Park, OH, (1973) p. 241–74.]

Mechanism. Calculations using molecular dynamics models have been made on models of dislocations, and a few on segments of grain boundaries. Fig. 6-10 shows several consecutive planes normal to the plane of a $\theta = 36.9°$ [100] tilt boundary ($\Sigma = 5$) in a bcc lattice. This boundary is 'high angle', but also has a relatively small repeat distance along the boundary in the planes shown. Thus the size of the model needed for calculations is reasonable. The results for diffusion in such a grain boundary indicate that both the motion energy and the formation energy of a vacancy are less than in the lattice, with the reduction in the energy of motion being about twice the reduction in the formation energy.[10] Other conclusions are:

- Vacancy jumps along the grain boundary core, of the sort $B \rightarrow D \rightarrow B$ or $B \rightarrow C \rightarrow B$ are the most frequent.

---

[10]R. W. Balluffi, in *Diffusion in Crystalline Solids*, ed. G. E. Murch, A. S. Nowick, Academic Press, 1984, p. 320–78.

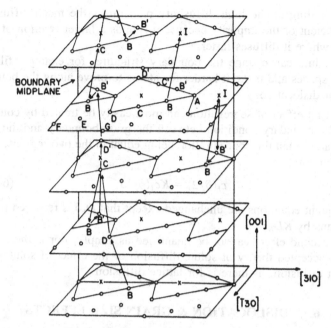

Fig. 6-10—Structure of $\Sigma R = 5$ [110] tilt boundary in bcc lattice showing repeated triangular units, and jump sequences along the core designated by arrows between atoms with letters beside them. (after R. W. Balluffi)

- Movement in the boundary normal to the core is much slower than along the core, that is, anisotropy in $D_b\delta$ persists at the high angles.
- Interstitials can form relatively easily, but move slowly.

The interatomic potentials used for such calculations do not give accurate values of the energies involved, but authors working with models of a variety of metals, for both dislocations and grain boundaries, all agree that diffusion is primarily by vacancy motion. This conclusion is further strengthened by the observation that the activation volume for diffusion in high-angle grain boundaries in silver is larger for boundary diffusion than for lattice diffusion, namely V_b is 1.1 Ω while $V_1 = 0.9$ Ω.[11] This large V_b also lends credence to the basic idea that vacancies exist in the boundary which are very similar to those existing in the lattice.

Alloying Effects. Alloying elements often segregate at grain boundaries or dislocations. This is especially true for elements whose solubility is low. This can have two distinctly different effects:

[11]G. Martin, D. A. Blackburn, Y. Adda, *Phys. Status Solidi*, 23 (1967) 223.

- The binding to the high diffusivity path raises the mean diffusion coefficient of the impurity because it spends a larger fraction of the time where it diffuses faster.
- The solute can change the boundary structure, for example fill in open spaces and make it harder for atoms to move along the boundary or dislocation.

The first effect of segregation can most easily be treated by considering the boundary condition between the grain boundary and the lattice. Rather than the concentration being equal in the two regions, Eq. (6-5) becomes

$$c_b(\delta/2) = Kc_1(\delta/2) \qquad (6\text{-}11)$$

Subsequent equations are unchanged except that $D_b\delta$ is replaced in all equations by $KD_b\delta$.

The second effect cannot be quantified as simply, nor is there any widely accepted theory of solute diffusion or the effect of solutes on solvent diffusion, as there is for lattice diffusion.

6.3 DISLOCATION & GRAIN SIZE EFFECTS

The treatment above dealt with the measurement of D_b and D_p in bicrystals. Consider now the effect of randomly oriented dislocations, or a fine grain size. These are of interest in studying the kinetics of diffusion limited processes at or below half the melting temperature, and are central in determining the rate of diffusion in fine grained thin films.

There is great similarity between the results for diffusion enhanced by a three dimensional array of dislocations[12] and that arising from a 3-D array of grain boundaries.[13] We will describe the results for dislocations, and then compare the results with those for boundaries. Consider the regular array of dislocation pipes of radius 'a' and separation $2Z$ as shown schematically in Fig. 6-11. They are normal to the free surface, which has solute on it. The section can show three different types of solute distributions depending on the ratio of the mean diffusion distance in the lattice $\sqrt{D_1 t}$ to the separation Z, and remembering that D_p is always much larger than D_1.

- $\sqrt{D_1 t} \gg Z$, an atom interacts with several dislocations in diffusing this far through the lattice. The dislocations increase the effective

[12]A. D. LeClaire, A. Rabinovitch, in *Diffusion in Crystalline Solids*, ed. G. Murch, A. Nowick, Academic Press (1984) pp. 259–319.

[13]D. Gupta, D. R. Campbell, P. S. Ho, in *Thin Films—Interdiffusion and Reactions*, eds. Poate, Tu & Mayer, J. Wiley & Sons, (1978) pp. 161–242.

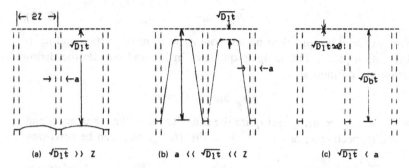

Fig. 6-11 — Schematic representation of the various concentration profiles that can develop with different ratios of $\sqrt{(D_l t)}/Z$. The solute source is at the top of each figure.

diffusivity D_{eff} for the solid, so that the penetration depth is greater than it would be without dislocations. The advancing isoconcentration lines are relatively flat near the dislocations. This will be referred to as 'A-kinetics.'

- $\sqrt{D_l t} \ll Z$, but $\sqrt{D_l t} \gg a$. The solute field around each dislocation develops independently of its neighbors, and the situation is similar to the case of grain boundary diffusion in bicrystals analyzed above. B-kinetics.

- $\sqrt{D_l t} < a$, Diffusion occurs only in the dislocation pipes with no loss to the surrounding lattice. This is rare in bulk samples, but in a thin film at $T = 0.5\,T_m$, D_p/D_l can equal 10^8 (see estimate for Ag above) and diffusion through a one micron thick film can occur without any solute being lost to the lattice. C-kinetics.

These three cases are next treated in more detail.

(a) Consider the dislocations randomly oriented with an average separation of $2Z$. If $\sqrt{D_l t} > 6Z$ then a diffusing atom encounters several dislocations over the time t, and the effect of the dislocations is to increase the average jump frequency of the atoms in an isotropic manner, thus $D_{eff}/D_l > 1$. If the heat flow analogy of the grain boundary diffusion problem was a system consisting of a sheet of aluminum foil between two sheets of plastic, then the heat flow analogy here would be a system consisting of fine aluminum wires randomly distributed in plastic. The relation between D_{eff} and D_l can be obtained by a random walk argument due to Hart.[14] Assume that each atom makes n jumps in a pure metal single crystal containing many randomly oriented dislocations. Ignoring correlation effects and assuming all jumps are of length r, the net displacement for each atom after n jumps is R_n and the average displacement

---

[14]E. Hart, *Acta Met.*, 5 (1957) 597.

$$\overline{R_n^2} = nr^2 \qquad (6\text{-}12)$$

Now of the n jumps taken by an atom n_1 were in the lattice and n_p in dislocation pipes. The jump frequencies inside and outside the dislocation are defined as

$$t_p \, \Gamma_p = n_p \text{ and } t_1 \, \Gamma_1 = n_1$$

where $n = n_p + n_1$, t_p and t_1 are the times spent in the dislocation and lattice respectively, and $t = t_p + t_1$. Eq. (6-12) can then be rewritten

$$\frac{\overline{R_n^2}}{t} = \Gamma_p \, r^2 \frac{t_p}{t} + \Gamma_1 \, r^2 \frac{t_1}{t} \qquad (6\text{-}13)$$

Except for a geometric constant, $\overline{R_n^2}/t$ equals D_{eff}; and except for the same constant, $\Gamma_p \, r^2$ and $\Gamma_1 \, r^2$ equal D_p and D_1, respectively. Thus D_{eff} for a single crystal is

$$D_{\text{eff}} = gD_p + D_1(1\text{-}g) \simeq D_1(1 + g(D_p/D_1)) \qquad (6\text{-}14)$$

where $g = t_p/t$. For a pure metal g is $(a/Z)^2$, or the fraction of atoms in the pipe. Some straightforward arithmetic indicates that if $D_p/D_1 = 10^7$ then the dislocations in an annealed crystal ($10^6/\text{cm}^2$) have a negligible effect on D_{eff}, but if the dislocation density is increased a hundred fold by slight cold working, D_{eff} is increased significantly. For the diffusion of a solute that tends segregate to the dislocation, $g = K(a/Z)^2$.

(b) In this case there is a region well below the surface in which the lattice around each dislocation pipe contains solute which entered by diffusion along the dislocation, that is $\sqrt{D_1 t} \gg a$. However, these solute fields do not overlap with one another because, $\sqrt{D_1 t} \ll Z$. Thus a semilogarithmic plot of the average concentration $\bar{c}(y)$ in a section at a depth y beneath the surface will show two regions. Near the surface where lattice diffusion is dominant $ln(\bar{c})$ is proportional to y^2 and the line has a slope $(-1/4D_1 t)$. Beneath this at a depth of say $y > 5\sqrt{D_1 t}$, $ln(\bar{c})$ is proportional to y (not $y^{6/5}$ as in the case of grain boundary diffusion) and the slope is given by

$$d\,ln(\bar{c})/dy = A/[a(D_p/D_1)^{1/2}] \quad \text{with } A \simeq 0.8 \qquad (6\text{-}15)$$

This slope is independent of time, in distinct contrast to the case for a tail due to grain boundary diffusion (see Eq. 6-6). Also, the slope is independent of: the density of dislocations, and whether the surface concentration is held constant or a thin film is applied. However, the intercept of a plot of $ln(\bar{c})$ vs. y extrapolated to $y = 0$ would depend on each of these variables.

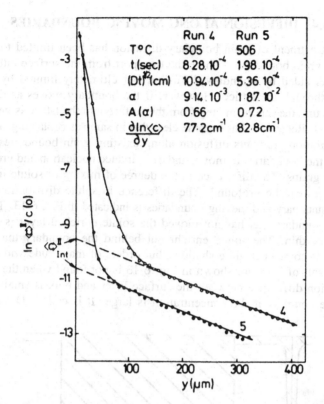

Fig. 6-12—Penetration curve for Na* into an NaCl single crystal showing a tail due to diffusion along dislocations. (Y. K. Ho, Thesis, Imperial College, London, 1982. See LeClaire & Rabinovitch)

Fig. 6-12 shows two penetration plots for a *Na* tracer diffusing from a thin surface film into pure NaCl single crystals in the low temperature (extrinsic) range. Note that the linear plot fits the data quite well, and that increasing the time by a factor of 4 does not change the slope. Both of these observations confirm that the tail is due to separate dislocations not dislocations aligned in low angle boundaries. Ho also treated the contribution of these dislocations to the electrical conductivity and found that

$$\sigma = (F^2D/RTf)\pi[a^2(D_p/D_1)]d \tag{6-16}$$

where d is the dislocation density. This adds one more mechanism to the list of possibilities that can influence σ and D in the Extrinsic range.

(c) Here the solute all stays in the dislocation pipes, and $c_p(y,t)$ is obtained in the same way used in Chap. 1 with D_p replacing D.

6.4 DIFFUSION ALONG MOVING BOUNDARIES

Our treatment of grain boundary diffusion has been limited to stationary grain boundaries. There the transport from the surface into the lattice is aided by boundary diffusion, but ultimately limited by diffusion through the lattice. However, if the boundary moves as diffusion occurs, the rate of mixing from the surface into the lattice is greatly increased. Recently it has become clear that in samples containing strong concentration gradients diffusion along existing grain boundaries can induce the boundaries to move, and even induce nucleation and growth of new grains. The difference in the degree of mixing of solute in the two cases can be profound. The difference in solute distribution between stationary and moving boundaries is indicated in Fig. 6-13. There at the boundary that has not moved the solute enriched layer is deep and very thin. The solute enrichment behind the boundary that has moved is somewhat more shallow, but orders of magnitude wider.

The sort of behavior shown in Fig. 6-13 is not found when the concentration difference between the surface layer and bulk is small, but is quite common if the concentration is large. It is called, Diffusion

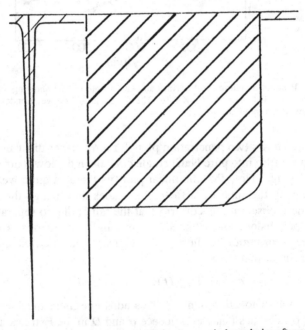

Fig. 6-13 — Cross section of sample containing two grain boundaries after exposure of the upper surface to a solute vapor. One boundary has exhibited DIGM along a portion of its length, one has not. The shaded area reflects the relative volume of solute enriched matrix around the two boundaries.

Fig. 6-14—Photomicrograph illustrating the surface relief accompanying the local zincification of an iron foil 40 microns thick by DIGM. [M. Hillert, G. Purdy, *Acta Met.* 26 (1978) 333.]

Induced Grain-boundary Migration (DIGM). One of the systems which easily exhibits DIGM is Fe-Zn. Fig. 6-14 shows the surface of a sample which has undergone DIGM. The regions high in Zn show up along the original boundaries in relief and contained 6-7% Zn through the entire thickness of the foil. The remaining part of the grains showed no increase in Zn aside from a surface layer about 0.1 um deep formed by lattice diffusion. The solute rich regions shown in Fig. 6-14 formed both by the movement of existing boundaries, and the nucleation and growth of new grains.

Questions concerning DIGM can be grouped in three categories: what happens, why, and when? The first has been touched on above, and will be described further below. As for 'why?', DIGM is driven by the free energy of mixing solute into the lattice. There is no general agreement on how this free energy couples to the boundary to move it, and indeed there may be several mechanisms operating at different times and in different systems.[15] The central point for our purpose is that DIGM can occur in alloy systems, and when it does, it will greatly increase the apparent D.

The question of when DIGM occurs has several types of answers:

- DIGM has been observed in essentially all binary alloys where it has been looked for. It has been reported in a few binary ceramic systems, but has not been found in most ceramic systems investigated.

[15]P. G. Shewmon, G. Meyrick, in *Interface Migration & Control of Microstructure*, ASM (1985) 7–17.

- DIGM has been reported to occur over a range of temperatures up to almost the melting point, but the effect is most pronounced and important at low temperatures.
- Only high angle boundaries exhibit DIGM. This may be due to their higher values of D_b, but probably has more to do with their ability to generate the many new sites needed to incorporate the solute into the alloy lattice.
- The movement is not continuous in time for a given boundary, but may stop and start again, may stop and then reverse, or may stop and not start again.
- The rate of both nucleation and growth increases as the Δc across the boundary increases.

6.5 SURFACE DIFFUSION AND SHAPE CHANGE

There are a variety of diffusion controlled phenomena familiar to a materials scientist which involve a change of shape and are driven by surface tension (surface free energy) or applied stresses too low to move dislocations. Examples are:

- creep at low stresses in polycrystalline materials can occur by diffusion through the lattice (Nabarro-Herring creep), or along grain boundaries (Coble creep).
- the sintering of powders occurs by the diffusion (lattice or boundary) from regions of sharp curvature to low curvature.

The principles which go into calculating the rate of each of these processes are understood. However, the actual calculation of the rate is often made difficult by the complicated geometry.

An example of surface tension driven diffusion with particularly simple geometry is the smoothing of a surface with a sinusoidal ripple in it. This problem is chosen because:

- A straight forward mathematical analysis exists.
- The analysis is borne out by experiment.
- The results provide a means for determining the surface diffusion coefficient.
- The analysis describes the relative contributions of competing transport processes.
- It provides a basis for treating other geometries by Fourier analysis.

Analysis of Surface Smoothing.[16] Consider the sinusoidally curved surface shown in Fig. 6-15 whose elevation y is given by the equation

---

[16]W. W. Mullins, "Solid Surface Morphologies Governed by Capillarity," *Metal Surfaces*, ASM, Metals Park OH (1963) p. 17–66.

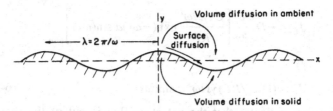

Fig. 6-15—Transport processes causing the decay of a sine wave in a solid surface. (W. W. Mullins)

$$y(x,t) = a(t) \sin(\omega x) \qquad (6\text{-}17)$$

The chemical potential of the atoms on the hills is higher ($\mu > 0$) than that beneath a flat surface ($\mu = 0$), while in the valleys the chemical potential is lower ($\mu < 0$). The magnitude of this variation is related to the surface tension γ and the curvature K by the equation

$$\mu = -\Omega\gamma K \qquad (6\text{-}18)$$

where Ω is the atomic volume. For a surface like this where the height y is a function only of x, $K = y''/(1 + y'^2)^{1/2}$ where y' and y'' are the first and second derivatives of y with respect to x. Thus

$$\mu(x,t) = -\Omega\gamma y'' = \Omega\gamma a(t)\omega^2 \sin(\omega x) \qquad (6\text{-}19)$$

where Eq. (6-17) is used to obtain the final equality. As indicated in Fig. 6-15 transport from the hills to the valleys can occur by three paths, diffusion through the gas, the lattice, or along the surface. If we assume that surface diffusion occurs in a thin layer of thickness δ, the flux crossing unit length on the surface is

$$J_s = (D_s\delta/RT)\partial\mu/\partial x \qquad (6\text{-}20)$$

where the derivative along the surface, $\partial/\partial s$, has been set equal to $\partial/\partial x$, an approximation good for small slopes. Combining Eqs. (6-19) and (6-20) gives

$$J_s = -(D_s\delta/RT)\Omega\gamma\, a(t)\omega^3 \cos(\omega x) \qquad (6\text{-}21)$$

The variation in chemical potential along the surface also gives rise to fluxes of atoms through the solid, and through the gas. The concentration variation in each phase can be approximated by the steady-state solution to the diffusion equation, $\nabla^2 c = 0$. There exists a solution $c(x,y)$ which satisfies Eq. (6-19) as a boundary condition along the surface and decays to a constant value far from the interface. If this is differentiated to get the flux normal to the surface the equation is

$$J_1 = -D_1 \frac{\partial}{\partial y} \left[\frac{C_o \gamma \Omega \omega^2}{RT} \exp(-\omega y) a \sin(\omega x) \right]_{y=0}$$

or

$$J_1 = (D_1 c_o / RT) \gamma \Omega \omega^3 a \sin(\omega x) \qquad (6\text{-}22)$$

This equation indicates that the net atom flux is out of the regions where $\sin(\omega x)$ is positive and into the regions where it is negative.[17] A similar equation holds for the gas phase, the only change being that $D_1 c_o$ is replaced by D and c for the gas phase. However, the vapor pressure is usually so low that transport through the gas phase is negligible, and it can be neglected.

The local rate of change of height of the surface at any point is given by the sum of the flux away from the surface into the solid and the divergence of the surface flux

$$\partial y / \partial t = -\Omega(\partial J_s / \partial x) - \Omega J_1 \qquad (6\text{-}23)$$

This is a conservation equation like Fick's Second Law, but the accumulation of the pure material appears as a shape change, not a change in concentration. Substituting Eqs. (6-21) and (6-22) into this equation and canceling $\sin \omega x$ gives

$$da / dt = -[B\omega^4 + C\omega^3] a \qquad (6\text{-}24)$$

with[18]

$$B = D_s \delta \gamma \Omega / RT \quad \text{and} \quad C = D_1 \gamma \Omega / RT$$

Since a pure solid is being considered, $c_o \Omega = 1$ has been used in the equation for C. Eq. (6-24) can be integrated to give $a(t)$. This can be used with Eq. (6-00). Since a pure solid is being considered, $c_o \Omega = 1$ has been used in the equation for C. Eq. (6-24) can be integrated to give $a(t)$. This can be used with Eq. (6-17) to give the complete profile, $y(x,t)$, but experimental observations are made on the decay of the amplitude with time. This equation is

$$d\ln a / dt = -(B\omega^4 + C\omega^3) = -B\omega^4 (1 + D_1 \lambda / 2\pi D_s \delta) \qquad (6\text{-}25)$$

where the last equality comes from replacing ω with the wave length λ using the equation $2\pi / \lambda = \omega$. Thus at small values of λ (large ω) the rate of decay in amplitude is controlled by surface diffusion and it varies as $D_s \delta / \lambda^4$, while at larger wavelengths volume diffusion is

---

[17]It is easier for some to visualize this flow of atoms through the solid by consider the equal and opposite flow of vacancies.

[18]Mullins' expression is $B = D_s \nu \gamma \Omega^2 / RT$. The difference is in his use of the term $\nu = \delta / \Omega$ where he calls ν "the number of atoms per unit area" on the surface.

rate controlling and the rate varies as D_l/λ^3. Surface and volume diffusion paths operate in parallel here. The thickness of the effective diffusion cross section in the lattice increases with λ while the thickness of the surface layer transporting material, δ, is a constant independent of wavelength.

Other Shape Change Techniques. The surface diffusion coefficient can be determined by following the rat of smoothing of a set of parallel grooves of spacing μ that have been etched into the surface. D_s has also been measured by following the development of other shape changes.[19] A more general equation for analyzing these changes is obtained by combining the flux equation (6-21) and the conservation equation to give the differential equation for pure surface diffusion

$$\partial y/\partial t = B\partial^4 y/\partial x^4 \qquad (6\text{-}26)$$

As an example this equation can be used to describe the development of a groove where a grain boundary meets an initially flat surface. Local reduction of the surface free energy gives an equilibrium groove angle which remains constant as the groove develops. Atoms move away from the sharply curved region near the groove root to form ridges on the relatively flat surface on either side. The largest and most easily measured dimension of a groove is the width between the tops of the ridges. The increase in this width w_s with time is given by

$$w_s = 4.6\,(Bt)^{1/4} \qquad (6\text{-}27)$$

Another example of a surface diffusion controlled process is the creep of a pure polycrystalline solid with an applied stress which is too low to move dislocations. The creep then occurs by the diffusion of atoms from grain boundaries with no normal stress to boundaries with a normal stress acting on them. The diffusion can occur by lattice diffusion, in which case it is called Nabarro-Herring creep, or by grain boundary diffusion, in which case it is called Coble creep. To make the calculation easier, consider the solid to be made of cubic grains of length L on a side as shown in Fig. 6-16. On boundaries normal to the applied load the chemical potential of atoms is reduced by $\delta\mu = -s\Omega$ due to the work (stress times displacement) that the applied stress, s, does if an atom is added to these boundaries, i.e. the sample is extended. There is no strain in the direction of the applied stress if an atom is added or removed from the vertical boundaries, i.e. the sample is not extended. Thus there will be a chemical potential difference and a diffusive flow through the lattice from one set of boundaries to the other.

[19]N. A. Gjostein, *Techniques of Metals Research*, vol. *IV part 2*, ed. R. Bunshah, Interscience Publ., (1970) pp. 405–57.

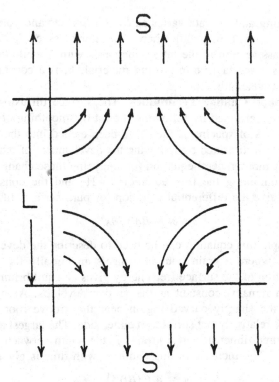

Fig. 6-16—Diagram showing flux of atoms inside a square grain under the applied stress s. The flow of vacancies will be opposite that shown by the curved arrows.

There is an equal and opposite flow of vacancies. At steady-state, local stresses will develop near the corners to make the flux into or out of the boundary equal at each point along the boundary. The average flux caused by the applied stress s is then

$$J_1 = -\frac{D_1}{RT\Omega}\frac{\delta\mu}{\delta y} \simeq \frac{D_1}{RT\Omega}\frac{s\Omega}{L/3} \tag{6-28}$$

where $L/3$ is taken to be the average diffusion distance between the boundaries under tension and those with zero normal stress. Now the rate at which a grain lengthens is $dL/dt = 2J_1\Omega$, so the overall strain rate caused by this shape change of the grains is

$$\frac{d\epsilon}{dt} = \frac{dL}{Ldt} = \frac{2J_1\Omega}{L} = \frac{nD_1s\Omega}{RTL^2} \tag{6-29}$$

where n is a number with a value between 10 and 40 depending on the grain shape, and stress state.[20] Note that the strain rate varies as the first power of the stress, and the reciprocal of the grain size squared. Both of these distinguish this type of creep from creep involving dislocation glide, or grain boundary diffusion.

If the transport occurs not by lattice diffusion but along the grain boundaries, the expression for the strain rate is $n(D_b\delta/LRT)(s\Omega/L^2)$. Adding this to Eq. (6-29), gives an equation for the strain rate when transport occurs by both paths

$$\frac{d\epsilon}{dt} = \frac{nD_b\delta s\Omega}{RTL^3}\left[1 + \frac{D_lL}{3D_b\delta}\right] \tag{6-30}$$

Note that the grain boundary contribution to the strain rate varies as $1/L^3$ so it will be more important at smaller grain sizes. Comparing Eqs. (6-30) and (6-26) one sees again the dimensionless variable $(D_b\delta/D_lL)$, which determines whether surface or lattice diffusion is controlling. It always appears in equations where the two paths work in parallel and indicates which process is dominant. It is an example of a general scaling law developed by Herring.[21]

Field Emission Spectroscopy. A quite different technique for studying surfaces involves field emission microscopy. In these instruments a very high electric field is applied to a sharply pointed metal wire (radius of curvature roughly 10-60 nm). In a high vacuum, the tip emits electrons in a pattern that reflects the atomic structure and composition of the surface (field electron microscopy). If a slight pressure of helium is admitted around the point, helium atoms are adsorbed on the surface of the point, ionized, and emitted from the surface. These emitted ions image the atomic details of the surface (field ion microscopy). The metal most commonly worked with is tungsten, primarily because it is strong enough to resists the high stresses developed in the tip by the electric field, but also because it can be easily cleaned by heating in a high vacuum. Two types of experiments have been performed using these techniques, one involving shape change and the other the diffusion of adsorbed atoms.

If the clean tip is heated slightly, the atoms rearrange themselves so the surface is bounded by flat, low index planes of the crystal. Using the atomic resolution of field ion microscopy the motion of individual atoms can be observed, and the diffusion coefficient of atoms across different faces and along different directions can be measured. This will be discussed in the next section.

A useful model for describing the structure of such surfaces is the

[20]W. Nix, *Metals Forum*, 4 (1981) 38–43.
[21]C. Herring, *J. Appl. Phys.*, 21 (1950) 301.

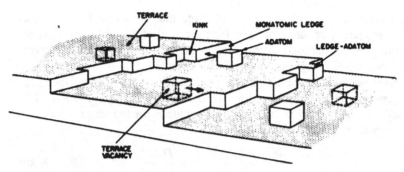

Fig. 6-17 — Terrace-ledge-kink model.

terrace-ledge-kink model shown in Fig. 6-17. If an imaginary plane is passed through a crystal near the orientation of a closely packed plane and all of the atoms above the plane removed, the surface of the crystal remaining will consist mostly of these closely packed planes. The edge of these planes will appear as steps, and have kinks in them. This is referred to as a ledge-step-kink model. At any finite temperature a few atoms will break away from the kinks and be mixed amongst the empty sites on the ledge as adsorbed atoms.

If the tip is heated to a higher temperature, though still well below half the melting point, atoms move from the very sharply curved tip to the less curved shank of the needle. This shape change is yet another example of surface curvature driven diffusion and D_s over a range of surface orientations is obtained from the rate of shrinking of the tip, dz/dt, by the equation[22]

$$dz/dt = 1.25 \, B/r3 \qquad (6\text{-}31)$$

where r is the radius of curvature of the tip, and B is defied in Eq. (6-24).

The movement of individual atoms on the surface can also be studied. With the atomic resolution of field ion microscopy, successive photos can be compared to indicate the motion of matrix atoms over a flat, low index plane. Or, with field electron microscopy the spreading of a solute layer can be followed. If a solute such as oxygen is admitted and adsorbed on one side of the point, it changes the work function locally, and the local intensity of the pattern of emitted electrons reflects this. The rate of spreading of the solute over the metal surface can then be followed by following the changing pattern. From such studies the variation of D with surface orientation, surface concentration, and temperature can be measured.

---

[22]F. A. Nichols, W. W. Mullins, *J. Appl. Phys.*, *36* (1965) 1826.

6.6 SURFACE DIFFUSION DATA AND MECHANISMS

Surface diffusion coefficient measurements are less accurate than those for lattice diffusion, the most prominent reason being the difficulty of controlling or specifying the wide range of surface structures sampled by different techniques. For example consider the three ways of measuring the surface self diffusion coefficient D_s:

- The blunting of a FIM tip removes hemispheres of atoms and the D_s measured is averaged over a wide range of orientations.
- The smoothing of a rippled surface measures D_s in a single direction, over a much narrower range of orientations, but again gives a value of D_s which is an average over many configurations, since entire planes of atoms are removed.
- FIM may be used to study the diffusion of a single adsorbed matrix atom on a surface. However, this is not D_s, but D_a for an adsorbed atom.

In addition to these structural variables there is another set of more chemical variables that arise from the adsorption of solute originally dissolved in the crystal, or from the ambient gas present to protect the surface. In closing the chapter, several examples of such measurements will be given.

Self Diffusion. The motion of individual atoms on a single surface can be followed with the FIM, and D_a for the adsorbed atom calculated from the mean displacement Δx seen after an anneal of time t through the equation $D_a = (\Delta x)^2/4t$. The high resolution of this technique requires that $\Delta x \simeq 1$ nm and thus the temperature must be low.

Note that the surface diffusion coefficient of adsorbed atoms D_a does not equal D_s, the diffusion coefficient that determines the rate of shape change discussed above. A very small fraction of the surface atoms, N_a, are adatoms at any instant, while D_s represents the average diffusion coefficient of all of the atoms in the surface layer. Thus the relation between the two can be represented by

$$D_s = D_a N_a \tag{6-32}$$

Representative data for D_a are shown in Table 6-2. The values of the preexponential constant D_o are small but consistent with those found for lattice diffusion with jumps to nearest neighbor sites and the entropy of motion, S_m, small but positive (see Chap. 2). Microscopic observations also indicate that the mechanism of diffusion of single atoms is by nearest neighbor jumps. The activation energy for diffusion on the close-packed (111) plane of the fcc rhodium is particularly low. (H_m in Table 6-2) As one moves to surface orientations farther away from close-packed planes, the density of steps increases to where

Table 6-2. D_s Parameters on Plane Surfaces[†]

Substrate	Atom	Plane	D_o (cm^2/s)	H_m (kJ/mol)
Single Atoms				
Rh(fcc)	Rh	(111)	2×10^{-4}	15
Rh	Rh	(110)[a]	3×10^{-1}	58
Rh	Rh	(331)[a]	1×10^{-2}	62
Rh	Rh	(100)	1×10^{-3}	85
W(bcc)	W	(110)	3×10^{-3}	87
W	W	(211)[a]	2×10^{-4}	72
W	Re	(211)[a]	2×10^{-4}	83
W	W	(111)	—	~172
Atom Pairs				
W	W	(211)[a]	7×10^{-4}	79
W	Re	(211)[a]	4.5×10^{-4}	75
W	Ir	(211)[a]	9×10^{-6}	65

[a]Motion along channels.
[†]G. Ehrlich, K. Stolt, in *Ann. Rev. Phys. Chem.*, *31* (1980) 603–37.

the surface consists only of steps. In the fcc lattice the (110) plane is such a surface, midway between two {111} planes, and the (331) is a similar corrugated plane midway between the {100} and {111}. Diffusion in such planes on Rh surfaces is observed only along the channels, cross-channel motion being much slower. However, on the (110) plane of fcc Pt D_a is more isotropic and it has been suggested that motion occurs by the ad-atom in one channel taking the place of an atom in the channel wall, and pushing that atom into an ad-atom position in the next channel; a model somewhat analogous to an interstitialcy mechanism inside a crystal. Such a diffusion mechanism has been observed for the motion of W and Ir on the (110) surfaces of fcc Ir.

Bcc tungsten is not close-packed and the activation energy for motion on the closest packed (110) is actually somewhat higher than that along channels of the corrugated (211). Again motion in the channels is much faster than cross-channel. One other mechanistic observation is that ad-atoms tend to form pairs, called dimers, and these pairs migrate without dissociation somewhat faster than single atoms. (Compare the data in Table 6-2 for single atoms of Re on W(211) and for pairs.)

Values of D_s obtained by shape change techniques average over many kinds of diffusive jumps and allow the measurement of D_s at much higher temperature. Fig. 6-18 gives results for nickel and tungsten.

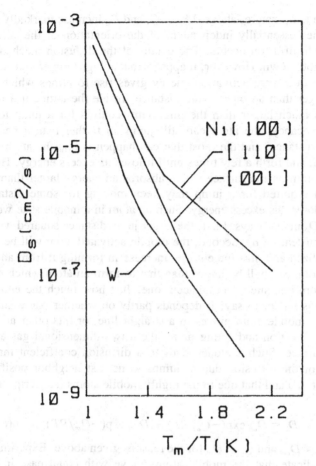

Fig. 6-18—D_s from tip blunting for W and from ripple smoothing in two directions on the (110) of Ni. [Vu Thien Binh, P. Melinon, *Surf. Sci.*, *161* (1985) 234–44.]

Clearly there is a change in slope in two of the curves. Such a change has been found for essentially all metals studied to date.

At low temperatures D has the following characteristics:

- D_o is similar to that found for lattice diffusion.
- Q_s is less than that for grain boundary diffusion (see Fig. 6-6), and there is appreciable variation of D_s with direction in a plane and between planes.
- Calculation of Q_a and D_o in the equation $D_a = D_o \exp(-Q_a/RT)$ for low index planes yields quite satisfactory agreement with experiment if a nearest neighbor jump model is assumed.[23]

[23]A. P. Voter, J. D. Doll, *J. Chem. Phys.*, *82* (1985) 80–92.

At high temperature, above $0.8T_m$, Q_s and D_o increase markedly and D_s becomes essentially independent of the orientation of the surface over which diffusion occurs. The details of the diffusion mechanism here are not known. However, it appears that at high temperature some process with a large activation energy gives rise to jumps which are much longer than an interatomic distance. Inside the lattice if a fluctuation is much larger than the minimum required for a jump to an adjacent vacancy, the atom can still jump no farther than a nearest neighbor distance, and the most that can happen is that the atom will jump back and forth a few times until it loses its excess energy. However, an atom on the surface which absorbs an energy larger than the minimum required for a jump may keep moving for some distance before it loses this excess energy. Such an atom in a mobile state would increase D_o in two ways. First, the larger jump distance squared would enter D_o instead of a_o^2. Second, the mobile activated atom will be less well localized and thus the entropy increase in forming it from an adsorbed atom, S_m, will be larger than that to form an atom which only moves from one site to an adjacent one. Just how much the entropy increases is harder to say. It depends partly on whether one assumes the highly mobile atom moves in a straight line, or hits other atoms, changing direction and acting more like a two-dimensional gas atom on the surface. Such a model leads to a diffusion coefficient that is the sum of the diffusion due to jumps to nearest neighbor positions (subscript 'a') and that due to the highly mobile atoms (subscript 'm'), or

$$D_s = D_{oa} \exp(-Q_a/RT) + D_{om} \exp(-Q_m/RT) \qquad (6\text{-}33)$$

Here $D_{oa} \ll D_{om}$ and $Q_a < Q_m$ for the reasons given above. Experiments on Cu indicate that the mobile atoms move with equal ease in any direction, for example across or along channels, but they are still bound to the surface since H_m is significantly less than the energy required to vaporize an atom. The average value of H_m that is measured could increase in temperature, as the mobile atoms dominating D_s become more energetic and move farther between activation and re-incorporation into the surface. Thus the dominant contribution to surface diffusion, as the melting point is approached, is by atoms which upon activation continue to move over the surface for 10 to 100 times the interatomic distance before they lose their exceptional kinetic energy and again come to rest in the surface.

Solute Effects. Two different phenomena come under the heading of solute effects, a study of the mobility of adsorbed solute atoms on surfaces, and the effect of adsorbed layers on the surface self diffusion of the substrate. The rate of spreading of adsorbed solute atoms across a surface has been studied for many years.[24] This work has often been done on tungsten surfaces with solute that change the emissivity in field emission spectroscopy. The technique does not allow atomic resolution, but can be done on clear surfaces as a function of surface concentration. The following general conclusions have been drawn:

- Diffusion of the adsorbed layer occurs at temperatures far below that required for desorption, thus the energy for motion is usually about 20% of the energy of desorption.
- D_s is often concentration dependent. On surfaces with two or three types of sites the first atoms diffusing onto a clean surface are trapped at the lowest energy sites. This can give results similar to that for hydrogen diffusion in steel with $D(c)$ increasing with concentration. In other systems adsorbed atoms interact to form dimers which have a higher mobility than individual atoms and a maximum in D_s occurs at an intermediate fraction of surface coverage.
- It is difficult to determine the value of D_s that would correspond to that for a tracer on a surface of constant composition since experiments always have a concentration gradient present and the thermodynamic factor $(d\kappa/dlnc_s)$ can make a substantial contribution to the observed value of D_s.

[24]G. Ehrlich, K. Stolt, *Ann. Rev. Phys. Chem.*, *31* (1980) 603–37.

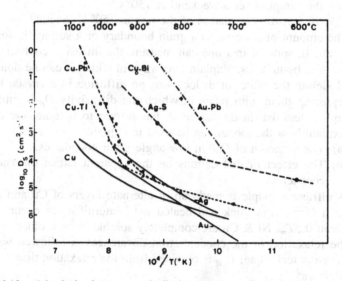

Fig. 6-19—Adsorbed solute can markedly increase D_s as shown by this comparison between D_s for pure Ag, Au, and Cu, and for surfaces with adsorbed layers. (F. Delamare G. E. Rhead)

Finally, we turn to the affect of adsorbed layers on surface self diffusion. For instance if initially clean surfaces of Cu or Ni samples containing carbon as an impurity are heated in a high vacuum, carbon diffuses from the interior of the sample, forms a carbon rich layer on the surface and greatly reduces the rate of surface self-diffusion. On the other hand sulfur on the surfaces of Cu or Ag can increase D_s by orders of magnitude. This effect is even more marked when vapors of Pb, Tl or Bi are maintained over the surface of these noble metals.[25] (see Fig. 6-19) Here D_s rises with the vapor pressure of the solute, and the effect is most marked at high temperature ($T > 0.8T_m$) where Q_s is large. Under these conditions the values of D_s and D_o are extraordinarily large. No satisfactory theory exists for these values of D_o and Q.

PROBLEMS

6-1. As a diffusion expert you are to calculate the thickness of Ag required to maintain at least a 99% Ag alloy on the surface for 5 years. The most accurate data you can find is a study of D for Ag in Cu between 750 and 1050° C. Extrapolating these data to 150° C, you find that a 1 μm layer of Ag will last for 100 years. A laboratory test shows that the silver layer completely diffuses into the sample over a weekend at 150° C.

Why was the calculation of the rate at 150° C invalid?

6-2. The amount of material in a grain boundary or a surface is very small. In spite of this one can measure the diffusion coefficient in these boundaries. Explain two ways in which it can be done.

6-3. Modeling the effect of dislocations on diffusion in a crystal by replacing them with pipes in which the diffusivity D_p is much higher than that in the lattice D_l has proven to be quite useful. Explain how the model can be used to explain:

(a) The variation of $D_b\delta$ in low angle grain boundaries.

(b) The effect of dislocations on the apparent lattice diffusion coefficient.

6-4. A diffusion couple is made up of alternate layers of Cu and Ni each $h = 4$ nm thick and heated until interdiffusion occurs at about $0.3T_m$. Ni & Cu are completely soluble in each other, and the relaxation of the concentration differences is followed with an x-ray technique. D_{eff} is obtained from the relaxation time $t_r = h^2/D_{eff}\pi^2$.

---

[25]F. Delamare, G. E. Rhead, *Surf. Sci.*, 28 (1971) 267.

(a) If the dislocation density of the films is $10^8/cm^2$ and the grain size is 50 nm, what is the ratio of D_{eff} to the lattice diffusivity D_1?

(b) Below what values of grain size would you expect D_{eff} to become much larger than D_1.

6-5. If the work is carefully done, D_T for Ag* can be measured in Ag at 540° K. If there are 10^6 dislocations/cm^2, what will be the ratio D_{eff}/D_1? Use data given in Table 6-1 to estimate D_p and list all assumptions made. (Assume the radius of the pipe is 0.2 nm)

6-6. Below 2/3 to 1/2 of the melting point, diffusion along dislocations makes a marked contribution to solute self-diffusion in very dilute alloy single crystals. It is quite probably that interstitials tend to segregate at dislocations, and also diffuse faster along dislocations than in the lattice. What justification is there for neglecting the effect of this enhanced diffusion along dislocations in deriving values for D from internal friction studies?

6-7. The grains of a metal can be idealized as a set of hexagonal grains all of the same diameter, L. If a small shear stress is placed on such a solid sliding occurs easily along the g.b., but the irregular shape of the grains dictates that the rate of shear is limited by the diffusion of material from regions in compression to those in tension, i.e. material must diffuse from one side of a grain to another. This diffusion can occur by both lattice and g.b. diffusion. Let $R = q_1/q_b$ be the ratio of the relative flows through the lattice and along the g.b.

(a) Will R increase or decrease with falling temperature? Why? Will R increase or decrease with falling grain diameter? Why?

(b) Write an equation relating R to D_1, D_b, L and g.b. thickness, δ. Assume that the stressed grain boundary can be approximated by a sine wave.

6-8. One of the ways used to study diffusion along isolated dislocations is to study the rate of diffusion through a single crystal sheet of thickness L which contains a regular array of parallel dislocations normal to the sheet. A layer of tracer is placed on one side and the rate of accumulation on the far side is observed. Assuming that $(D_1 t)^{1/2} < a_o$ give equations for the following:

(a) The delay time between the start of diffusion into a solute free sheet and the first appearance of solute on the far side.

(b) The initial rise with time of the total amount of solute on the far side assuming that the rate of surface spreading is so fast that the far surface is homogeneous and the rate limiting process is diffusion along the pipes. (The density of dislocations is d per unit area.)

Answers to Selected Problems

6-5. (a) The tracer distribution in the pipes is initially given by an error fcn. solution. Assuming detection when the far side concentration at the end of the pipe reaches 1% that of the source side, $t_d \simeq L^2/3.3D_p$.

(b) Activity $= (h^2/d^2)D_p t$.

7

THERMO- AND ELECTRO-
TRANSPORT IN SOLIDS

If a current of electricity, or a flux of heat, is passed through an initially homogeneous alloy an unmixing occurs, that is a concentration gradient develops. These effects are called electro-transport and thermo-transport, respectively. In electro-transport the atomic redistribution is similar to that studied in ionic conductors. However, in metals electrons carry essentially all of the electric current, and the ratio of electron to atomic currents is high. It appears that most of the atomic transport results from the impact of the large flux of electrons on the solute atoms making diffusive jumps. In thermo-transport, the redistribution of solute which occurs is analogous to the more widely studied thermoelectric effects that arise from the redistribution of electrons in a solid in a temperature gradient. The origin of the force driving the atoms is not clear.

Since a current or heat flow leads to the unmixing of an initially homogeneous single phase alloy, one must add terms to the flux equation that reflect these forces. Using the format of phenomenological equations (Chap. 4) the flux equation for an interstitial solute in an alloy would be

$$J_1 = -L_{11}(\partial\mu_1/\partial x)_T - L_{1q}(\partial T/\partial x) - L_{1e}(\partial\phi/\partial x) \qquad (7\text{-}1)$$

In what follows we consider the physical models and experimental data for the coefficients L_{1e} and L_{1q}. The treatment in each case will deal first with the diffusion of interstitial solute and then go on to treat mass transport in substitutional alloys.

223

7-1.   ELECTRO-TRANSPORT[1,2]

The flux equation for an interstitial can be written

$$J_i = \frac{-D_i N_i}{RT}\left(\frac{RT\partial \ln N_i}{\partial x} + F z_i^* E\right) \qquad (7\text{-}2)$$

where z_i^* is the effective charge on the interstitial, F is Faraday's constant, and E is the electric field (voltage gradient). This is the same equation used to describe diffusion and electrical conductivity in an ionic conductor, but here the processes that occur are different. For example, if a direct current is passed through a dilute Cu-H alloy, it is found that rather than hydrogen migrating as if it had a charge of $+1$, it migrates as if it had a charge z_i^* of -15 to -20. Thus the sign of z_i^* is the opposite of what one would expect from chemical valence, and the magnitude is much larger. In addition to an electrostatic force tending to make the hydrogen migrate as if it had a charge of $+1$ in the lattice (its true charge z_i), there is clearly a much larger force driving the atoms in the same direction as the electrons carrying the current in the Cu.

The generally accepted theory for this is that an atom in an activated state disrupts the flow of electrons more than an atom on a lattice site. The moving electrons which carry the electric current in a metal hit the atoms in activated states and in bouncing off them transfer momentum that biases the diffusive jumps in the direction the conduction electrons are moving. They would also inhibit jumps in the opposite direction. This force is termed the 'electron wind,' and calculations indicate it can be much stronger than the electrostatic force exerted by the field on the charged interstitial atom.

The force due to the electron wind can be calculated as follows. The number of collisions per unit time between the electrons and a moving atom is the product of the density of moving electrons n_e, their average velocity v_e, and the atom's cross section for collision with the electrons σ_e. For each collision the electron transfers on average the momentum that it has acquired during one relaxation time between collisions, $eE\tau_e$. Thus the rate of momentum transfer is $eE\tau_e(v_e n_e \sigma_e)$. This equals the force on the atom by Newton's First Law of motion. Thus z_i^* is the sum of the true charge z_i, and a 'wind' term of $\tau_e v_e n_e \sigma_e$. To simplify the equation, note that $l_e = \tau_e v_e$ is the mean-free-path of the electrons. Also, in some metals the current is carried by the motion of holes as

---

[1] H. B. Huntington, in *Diffusion in Solids*, ed. A. S. Nowick, J. J. Burton, Academic Press (1975), 303–52.
[2] For the most complete survey of the literature see Hans Wever, *Elektro- u. Thermotransport in Metallen*, J. A. Barth, Leipzig (1973). (in German)

well as electrons, so a comparable term $n_h l_h \sigma_h$ is added for holes

$$z_i^* = z_i + n_h l_h \sigma_h - n_e l_e \sigma_e \qquad (7\text{-}3)$$

In the transition elements electron holes play a dominant role in electrical conduction, and the sign of z_i^* changes for these elements. Figure 7-1 summarizes the data available for part of the Periodic Table. In the box for each element the position of the chemical symbol for the solute elements indicates whether the element migrates toward the anode ($z_i^* < 0$) or cathode ($z_i^* > 0$). The sign of the Hall Effect indicates the sign of the dominant charge carrier. It is negative for predominantly electron transport of current and positive for transport by holes. As an example, for Vanadium, which has a positive Hall Effect sign, hydrogen, carbon, nitrogen, oxygen, and vanadium atoms all are driven toward the cathode (+ electrode) by an electrical current.

Two generalizations can be made from the data in Fig. 7-1:

• For a given element most or all of the solutes, and the solvent itself, move toward the same electrode.
• The direction of motion correlates well with the column in the Periodic Table, and thus with the nature of the carrier, while it does not change with the nature of the impurity. Thus the electron wind model is in general agreement with experimental results.

The most accurate way to determine z_i^* for interstitial impurities is to pass a current through an isothermal sample until a steady-state is reached, and then determine the concentration gradient, $dln(c_i)/dx$, which is equal to $-z_i^* E/RT$ (see Eq. 7-2). Much of the accuracy stems from the fact that the determination of z_i^* requires no knowledge of the diffusion coefficient. The diffusivity only influences the time to achieve steady-state. The work on electrotransport of interstitials has often been

```
|   IVb   |    Vb    |   VIb   |  VIIb  |          VIIIb           |   Ib   |
|A      C|A       C|A      C|A     C|A      C|A      C|A      C|A      C|
|        H|        H|        |        |       H|        |     H|H        |
|   Ti   C|   V    C|  Cr    |  Mn   |  Fe   C|  Co   C|  Ni   C|  Cu     |
|N  (n)   |    (p) N|   (p)  |  (p)  |  (p)  N|  (p)   |  (n)  N|  (n)    |
|O        |    O    |        |        |   O    |        |   O    |         |
|Δ        |    Δ    |        |        |   Δ    |      Δ|Δ     Δ|         |
|        H|        H|        |        |        |        |     H|H        |
|C  Zr    |  Nb   C|  Mo   C|  Tc   |  Ru    |  Rh    |  Pd    |  Ag     |
|N  (p)   |    (p) N|   (p) N|       |  (p)   |  (p)   |  (n)   |  (n)    |
|O        |    O    |   O    |        |        |        |        |         |
|         |    Δ    |   Δ    |        |        |        |     Δ|         |
```

Fig. 7-1—Direction of electrotransport of interstitials. Transport to A:anode, or C:cathode. C,H,N,O: chemical symbol for interstitial, Δ:self transport. (n) negative, and (p) positive Hall Effect. [H. Wever in, *Electro- & Thermo-transport in Metals & Alloy*, ed. R. E. Hummel, H. B. Huntington, AIME, New York (1977) p. 37–52.]

motivated by the desire to use it as a means to purify reactive metals like transition metals, rare earths and actinides.[3] These metals have a high affinity for C, N, and O, but electromigration sweeps all of the interstitials in the same direction so all can be removed at once. Most of these metals have been purified to a higher degree with electromigration than by any other technique. The effectiveness of the technique also stems from the exponential increase in concentration at the end where the interstitials are swept.

For elements like hydrogen which exchange easily with the surrounding atmosphere, another techniques is used to measure z_i^*. Sensitive means for measuring the direction and the steady-state flow rate of the gas through a metal membrane are available, and these allow quite accurate measurements of the flux, and from this the quantity $D_H z_H^*$.[4]

Pure Metals. Electrotransport also occurs in pure metals. To measure z^* one must measure the net drift of matrix atoms (or equivalently vacancies) relative to the lattice. This has been done in two types of experiments. Both require inert markers in the sample. In one the drift of a tracer relative to the markers is measured, and in the other the divergence of the flow in a temperature gradient.

In the first type of experiment the end of a metal rod is coated with a thin layer of a radioactive isotope and inert (though possibly radioactive) particles that serves as markers. This rod is then butt welded to another rod of the metal so that the markers and tracer are in the middle of the sample. A direct current is passed through the welded rod, which is kept as close to isothermal as possible. Under these conditions, two types of atomic motion occur. The tracer spreads out by lattice diffusion down a concentration gradient, and all of the atoms drift in the same direction with a velocity v relative to the inert markers, under the influence of the electric current. The diffusion equation in this case is

$$\frac{\partial c_B}{\partial t} = \frac{\partial}{\partial x}\left(\frac{D^* \partial c_B}{\partial x} - n_B v\right) \tag{7-3}$$

where $v = $ (force)(mobility) $= (F z_i^* E)(D/RT)$. If we substitute $x - vt$ for x then $c_B(x,t)$ is replaced by $c_B'(x - vt, t)$ and the diffusion equation becomes

$$\partial c_B'/\partial t = D^* \partial^2 c_B'/\partial x^2 \tag{7-4}$$

---

[3]D. T. Carlson, in *Electro- & Thermo-transport in Metals & Alloys*, eds. R. E. Hummel, H. B. Huntington, AIME, New York, (1977) pp. 54–67.
[4]R. A. Oriani, O. D. Gonzales, *Met. Trans., 1* (1967) 1041.

with the solution

$$c_B'(x,t) = \frac{N}{(\pi D^* t)^{1/2}} \exp\left[-\frac{(x - vt)^2}{4D^* t} \right] \qquad (7\text{-}5)$$

where N is the quantity of material initially placed at the interface. Thus the tracer distribution is unchanged from the case without electrotransport, but is shifted from its initial position by a distance vt. Serial sectioning of the sample after an anneal allows the determination of vt as the distance between the maximum in c_B and the position of the inert particles. The value of D^* can be obtained from the distribution of the tracer. Figure 7-2 shows this experiment schematically. Note that $D = D^*/f$ appears in the equation for the velocity v while the 'correlated' D^* is obtained from the tracer distribution. It is always the case that D not D^* is the relevant diffusion coefficient to use when all the atoms are moving, e.g. in diffusional creep as discussed in Chap. 6.[5]

The other technique for determining the atomic flux relative to the lattice (and thus z^*) measures the dimensional changes of the sample during electromigration in a temperature gradient. Fig. 7-3 shows a sample heated by the direct current and cooled at both ends. The markers are scratches or indentations. The expansion or contraction of the region between two markers is given by the difference between the flux into and the flux out of the region. Consider the region between the center and the cold end of the sample. The flux is zero at the cold end, and if the cross sectional area of the sample stays uniform along

[5]H. B. Huntington, in *Diffusion in Solids*, ed. A. S. Nowick, J. J. Burton, Academic Press (1975), 306–12.

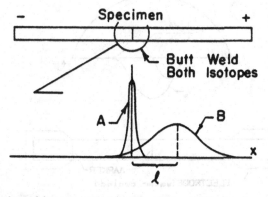

Fig. 7-2—Isothermal isotope method of determining z^*. The inert marker activity is A, and the matrix isotope is B. $vt = l$ is the displacement. (after Huntington, 1975)

its length, the velocity of the marker at the center equals the flux of atoms across the marker times the atomic volume Ω

$$v_m = J_A\Omega = -(D_A/RT)(Fz_A^*E)\Omega \qquad (7\text{-}6)$$

Here D_A is D_A^*/f or the uncorrelated D. It is found experimentally that the cross section of the region changes as well as the length. The product of $J_A\Omega$ and the cross sectional area at the marker thus equals the change of volume of the entire region, and the determination of z_A^* requires careful measurement of the variation of the cross section as well as the motion of the central marker. Fig. 7-3 indicates schematically how v_m and dA/dt might vary with position along the heated bar.

Table 7-1 summarizes data for many metals, clustered in groups that fall in the same column of the Periodic Table. Again the magnitude and sign of z^* suggest that the electron wind is the dominant force. The only elements which have positive values of z^* are transition elements with complex band structures such as Fe, Co, Pt and Zr. In these elements hole conduction plays an important role as indicated by the positive Hall Effect (see Fig. 7-1).

Thin Films. One of the commercial problems that has driven work on electrotransport stems from the failure of narrow thin-film conductors in printed circuits. These devices operate with direct current and the micron sized conductors carry current densities of over 10^6 A/cm^2. Failure frequently arises from the development of pores in the metal, followed by local heating and failure. A considerable effort has been devoted to understanding, and eliminating, this problem.

The first thing to note is that the conductors operate at below $0.4T_m$

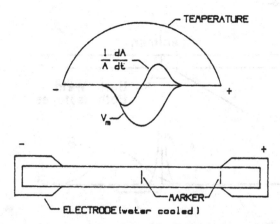

Fig. 7-3 — Vacancy Flux method showing markers and temperature distribution caused by heat generated by the current.

Table 7-1. z^*/f **for Pure Metals**[†]

Metal	z^*/f	Metal	z^*/f
Au	−8	Ni	−3.5
Ag	−9	α-Fe	2
Cu	−5	Co	1.6
		Pt	0.28
Li	−2		
Na	−3	β-Zr	0.3
		γ-U	−1.6
Zn ‖c	−2.5		
Zn ‖a	−5.5	Pb	$−47/f$
Cd ‖c	−2.0	Sn	−18
Cd ‖a	−4.1		
Al	−30 to −12		
In	−11.4		
Tl	−4		

[†]Data from H. Huntington, *Diffusion in Solids*, ed. A. S. Nowick, J. J. Burton, Academic press (1975), p. 329.

for aluminum. Thus essentially all of the transport is by grain boundary diffusion, not through the lattice. This conclusion is also born out by the fact that alloying to reduce the rate of diffusion through the lattice does not result in reduced failure rates, while alloying to reduce boundary diffusion leads to longer life.

If matter transport is predominantly along grain boundaries then anything that leads to a local divergence in the boundary flow can lead to the accumulation of vacancies (pores) or atoms (hillocks). Indeed pores and hillocks are found on grain boundaries in failed elements examined with the electron microscope. This flow divergence can arise from any of several sources, for example D_b may change due to a change in temperature, boundary structure, or boundary composition. It may also be caused by a change in boundary geometry. Figure 7-4 shows examples of two situations in which divergences can occur. In (a) the divergence is due to a sudden increase in grain size. The change in grain size leads to a grain boundary flux that has no place to go when the grain boundary ends on a larger grain. In (b) the divergence is due to a change in D_b stemming from composition change formed by evaporating a thin film of copper on the aluminum, and homogenizing at an elevated temperature to form a band of an Al-2%Cu alloy in the Al conductor.

The reason Cu reduces the grain boundary flux is less clear. A plausible model postulates that D in the grain boundary of any metal is much greater than D in the lattice because the density of relatively

Diffusion in Solids

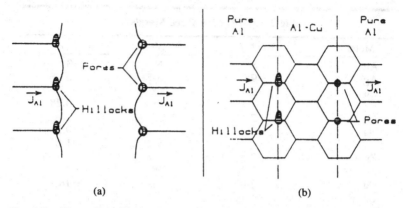

(a)   (b)

Fig. 7-4—Illustration of formation of pores and hillocks due to flux divergence in the grain boundaries of thin films carrying DC electricity. (a) Large change in grain size. (b) Change in composition changes D_b of boundary.

open sites (psuedovacancies) in the boundary is much greater than the equilibrium vacancy concentration in the lattice. The copper is thought to segregate to the boundary and fill many of these psuedovacancies, making them unavailable to aid the diffusion of Al along the boundary. Thus decorating the grain boundary with copper reduces D_b and the flux of aluminum relative to that in pure Al.

7-2.   THERMO-TRANSPORT—Interstitial Alloys

The fact that a temperature gradient can lead to the unmixing of an initially homogeneous alloy indicates a biasing of the jumps either up, or down, the temperature gradient. For a situation in which only one component is diffusing, e.g. an interstitial alloy, the flux equation can be written

$$J_1 = \frac{-D_1 N_1}{RT}\left[\frac{RT\partial lnN_1}{\partial x} - \frac{Q_1^*}{T}\frac{\partial T}{\partial x}\right] \tag{7-7}$$

D_1/RT is the mobility of component 1, $(Q_1^*/T)dT/dx$ is the effective force exerted by the temperature gradient, and $RT(\partial lnN_1/\partial x)_T$ is the chemical potential gradient at constant temperature. Q_1^* is the experimentally determined parameter which describes the sign and magnitude of the thermo-transport effect. It is called the heat of transport of component 1; the rest of our discussion of thermo-transport will be devoted to its interpretation and measurement. Note that D_1 is the isothermal diffusion coefficient. The temperature gradient changes neither the jump mechanism, nor the mean jump frequency at any given temperature; it does bias the direction of jumps.

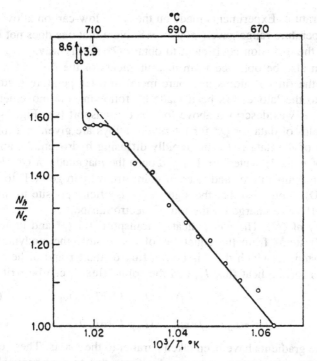

Fig. 7-5—Plot of carbon content (logarithmic scale) vs. $1/T$ for α-Fe annealed in a temperature gradient until steady state is attained. [P. Shewmon *Acta Met.*, 8 (1960) 606.]

Q^* can be measure either by letting the flux go to zero in a temperature gradient and measuring the concentration gradient at $J = 0$, or by measuring the flux through an open system under a known temperature gradient. Figure 7-5 shows the concentration gradient obtained when an initially single-phase iron-carbon alloy (0.01%C) was held in a temperature gradient until a steady state was established. The flux was then equal to zero and Eq. (7-7) gives

$$\frac{dlnN_1}{dx} = \frac{-Q_1^*}{RT^2}\frac{dT}{dx} \qquad (7\text{-}8)$$

During the anneal the carbon concentration became higher at the hot end, indicating that Q^* is negative.[6] The line drawn through the points gives a value of $Q_c^*(\alpha) = -96$ kJ/mol, and suggests it is independent

---

[6]The off-scale points on the left side of Fig. 7-5 result from the precipitation of iron carbide in this region and in no way affects the determination of $dlnN_1/dT$ in the single phase portion of the specimen. The redistribution of a second phase during annealing in a temperature gradient is discussed below.

of temperature. Experiments made on the same low-carbon alloy in γ-iron temperature range gave $Q_c^*(\gamma) = -8$. Note that one does not have to know the diffusion coefficient to obtain Q^* in this way.

Q^* can also be obtained from measurements of the flux. Measurement of the flux of atoms in a pure metal in a temperature gradient, relative to the lattice, has been made by following the movement of markers, as was described above for electrotransport (see Fig. 7-3).

Examples of data for Q^* for interstitial alloys are given in Table 7-2. Much of the data is for the rapidly diffusing hydrogen, H, and its isotope of mass 2, deuterium. For hydrogen the magnitude of Q^* clearly rises with temperature, and rises in going from hydrogen (H) to deuterium (D). Note also that the sign of Q^* is often opposite to that of z^*, the effective charge on the ion in electrotransport.

Theory of Q^*. The name 'heat of transport' for Q^* and its use in Eq. (7-7) stems from the equations of irreversible thermodynamics. For a system in which there is both a flux of matter and of heat, the equations for the heat flux J_q, and the solute flux J_1 can be written

$$J_1 = -L_{11}(\partial\mu_1/\partial x)_T - (L_{1q}/T)(\partial T/\partial x)_{N1} \qquad (7-9)$$

$$J_q = -L_{q1}(\partial l_1/\partial x)_T - (L_{qq}/T)(\partial T/\partial x)_{N1} \qquad (7-10)$$

where the gradients have been taken parallel to the x-axis. These equations express the flux as the sum of a force due to the chemical potential gradient at constant temperature and a temperature gradient at constant composition. The flux equation for J_1 can be rewritten

$$J_1 = -L_{11}[(\partial\mu_1/\partial x)_T - (L_{1q}/L_{11}T)(\partial T/\partial x)]$$

Comparing this with Eq. (7-7), if L_{11} is set equal to D_1N_1/RT then Q_1^*

Table 7-2. Q^* for Interstitial Alloys

Solvent	Solute	Q^* (kJ)	Ref.	Solvent	Solute	Q^* (kJ)	Ref.
bcc				fcc			
α-Fe	H	−23.5	1	Ni	H	−0.84	1
α-Fe	D	−22	1	Ni	D	−3.4	1
α-Fe	C	−59	2	Ni	C	−12.2	4
V	H	7.5	3	Co	C	6.3	2
V	D	20	3	Pd	H	6.3	2
V	C	−20.5	2	Pd	C	35	2

[1] O. Gonzales, R. Oriani, TAIME, 223 (1965) 187.
[2] J. Shaw, W. Oates, Met. Trans., 2A (1971) 2127.
[3] D. Peterson, M. Smith, Met. Trans., 13A (1982) 821.
[4] I. Okafor, O. N. Carlson, D. Martin, Met. Trans., 13A (1982) 1713.

$= L_{1q}/L_{11}$. But if $(\partial T/\partial x) = 0$, the ratio of Eqs. (7-10) and (7-9) gives

$$(J_q/J_1)_T = L_{1q}/L_{11} = L_{q1}/L_{11} = Q_1^* \qquad (7\text{-}11)$$

where the Onsager reciprocal relation, $L_{1q} = L_{q1}$, holds since we have selected the forces in Eq. (7-9) and (7-10) appropriately.[7] Q_1^* is thus the heat flux per unit flux of component 1 in the absence of a temperature gradient. If $Q_1^* > 0$, a heat flux parallel to J_1 will be generated by a solute flux; that is, to keep the region gaining solute atoms isothermal, heat must be removed from it. If $Q_1^* < 0$, J_q and J_1 are in opposite directions, and the region gaining solute atoms must receive heat to keep it isothermal.

Heat in a metal is carried by free electrons and elastic waves in the lattice (called 'phonons'). In the presence of a temperature gradient the jumps of the moving atoms are biased by the interaction of the atom with electrons, phonons, and/or gradient related assymetries in the activation process. The interaction is difficult to treat with precision, but qualitatively there are two contributions. One force stems from the interaction with flowing electrons, and the other from gradient related assymetries in the activation process.

In a simple metal in a temperature gradient the kinetic energy of the free electrons on the hot side of the sample are raised by adsorbing heat from the heat source, while on the cold side the kinetic energy of the free electrons is lowered by giving up heat to the heat sink. The heat is carried in the solid by the flow of more energetic electrons from the hot to the cold side, while less energetic electrons flow in the reverse direction to maintain charge neutrality. This gradient induced flow of higher energy electrons biases the jumps of the atoms in the same direction, that is it induces an electron 'breeze' in the same way that a current flow does in electro-transport. Qualitatively this predicts that if z^* is negative then Q^* will be positive, since the flow of negatively charged electrons down the temperature gradient biases the motion of atoms in the same direction as that of the electrons. Such a correlation between the sign of $-z^*$ and Q^* is common, but not universal.

A kinetic argument due to Wirtz treats the biasing of jumps without any reference to the flux of energy through the solid. In an isothermal system the probability that a given solute atom will be in the high-energy configuration required for a jump is related to the temperature of the region through the factor $\exp(-H_m/RT)$. In the presence of a

---

[7]A useful, simple introduction to these relations is given by K. G. Denbigh, *The Thermodynamics of the Steady State*, Methuen & Co., (1951). The forces have been chosen so that the product of the flux J_i and the force X_i has the units (temperature)(entropy)/(time)(volume).

temperature gradient, the average temperatures of the original, intermediate, and final planes of the jumping solute will be different. As a result, the frequency with which the required high-energy configuration is established for a jump to the higher temperature side may differ infinitesimally from that for a jump to the lower temperature side. To obtain an equation for these two jump frequencies, assume that H_m consists of three parts:

1. That given to the atoms on the original plane of the solute (H_o),
2. That given to the atoms in the intermediate plane of the jump (H_i),
3. That required to prepare the final plane for the jumping atom (H_f).

The jump frequency for the atoms moving up the temperature gradient will then be proportional to the product

$$\exp\left(\frac{-H_o}{RT}\right) \exp\left(\frac{-H_i}{R(T + \Delta T/2)}\right) \exp\left(\frac{-H_f}{R(T + \Delta T)}\right)$$

where the average temperature difference between the original and the final plane of the jumping solute is ΔT. The jump frequency for atoms jumping in the reverse direction between the same two planes will be proportional to:

$$\exp\left(\frac{-H_f}{RT}\right) \exp\left(\frac{-H_i}{R(T + \Delta T/2)}\right) \exp\left(\frac{-H_o}{R(T + \Delta T)}\right)$$

The middle term in both of these equation is the same so the ratio of the jump frequencies is

$$\exp\left(\frac{-H_o + H_f}{RT}\right) \exp\left(\frac{-H_f + H_o}{R(T + \Delta T)}\right)$$

If n_h and n_c are the number of atoms per unit area on the hotter and colder planes, respectively, the condition for zero flux between the planes is

$$\frac{n_h}{n_c} = \exp\left(\frac{-H_o + H_f}{RT}\right) \exp\left(\frac{-H_f + H_o}{R(T + \Delta T)}\right)$$

$$= \exp\left(\frac{-(H_o - H_f)\Delta T}{RT^2 (1 + \Delta T/T)}\right) \tag{7-12}$$

But $(n_h - n_c)/n_c$, and $\Delta T/T$ are both much less than one, so this equation can be put in differential form as

$$\frac{d \ln(n)}{dx} = \frac{-(H_o - H_f)}{RT^2} \frac{dT}{dx} \tag{7-13}$$

Comparing this with Eq. (7-8), the equation for Q_1^* for this model is

$$Q^* = H_o - H_f \tag{7-14}$$

Since H_o, H_i, and H_f were defined such that $H_m = H_o + H_i + H_f$, Eq. (7-14) requires that $|Q^*| \leq H_m$ but allows any value of Q^* in this range.

This model indicates that Q^* is influenced by the spatial distribution of the activation energy required for a jump. In Chap. 2 the atomic mechanisms and atomic rearrangements required for an atomic jump were discussed. The most common situation was one in which the diffusing atom had to pass through a constriction on its way to a relatively open, new site. If the primary barrier to diffusion were the movement of constricting atoms out of the way so that the diffusing atom could pass, then most of H_m would be located in the intermediate plan, H_m would then be about equal to H_i, and Eq. (7-14) indicates that Q^* would be almost zero. On the other hand, if the main part of H_m were required to make the diffusing atom execute violent enough oscillations to move it to the saddle point (the constriction always being relatively open), the result would be $H_m = H_o$, and $Q^* = H_o$. In this latter case the solute would tend to concentrate at the cold end. This result can be seen by noting that if $H_m = H_o$, the jump frequency of the solute atoms on the hotter of two adjacent planes will always be greater than that of the solute on the colder plane. Thus if the number of atoms jumping from the cold plane to the hot plane per unit time is to equal the number making the reverse jump, there must be more solute atoms on the colder plane.

A system to which the Wirtz model seems to apply is carbon in iron. The data in Table 7-2 indicate that Q^* in bcc iron is negative and roughly equal in magnitude to the H_m, while in fcc iron Q^* is close to zero even though H_m is almost twice that in bcc iron. In an fcc lattice an interstitial atom must pass through a pronounced constriction in a jump from one interstitial site to another (see Fig. 2-3). Thus it is plausible to say that H_i would be an appreciable part of H_m, and Q^* would be expected to be small, as it is. The large negative value of Q_c^* in the bcc α-iron may at first seem anomalous since its interpretation using Eq. (7-14) requires that most of H_m goes into preparing the plane of the final site for the jump. However, examination of the bcc lattice, e.g. Fig. 7-6, shows that the moving atom must pass through no constriction midway along its jump. The major barrier to be overcome in the movement of an interstitial atom from one site to an adjoining one appears to be the moving apart of two iron atoms in the final plane so that the carbon atom can jump from its initial position into a position between them.

Discussions of the theory of Q^* usually involve some combination

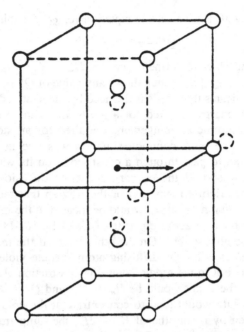

Fig. 7-6—Model of interstitial atom moving in bcc lattice. The dashed circles represent the new positions of matrix atoms after the interstitial atom makes the jump shown by the arrow.

of these two effects, that is the 'electron breeze', and the Wirtz model. The theory is qualitatively satisfactory, but quantitatively poor, since there is no way to determine the exact contribution of each effect to Q^*.

Precipitation and Phase Redistribution. Interstitials diffuse rapidly and also often form precipitates with a low solubility. This can cause the precipitates in a two phase alloy to redistribute in a temperature gradient. Or, it can lead to the formation of a precipitate at one end of a sample in an alloy which was a solid solution when the alloy was placed in a temperature gradient. As an example of the latter case, consider the diffusion of hydrogen in zirconium. Q^* is positive for this system (6 kcal/mol), thus thermo-transport pushes the hydrogen toward the cold end of a sample in a temperature gradient. The solubility of zirconium hydride drops as the temperature decreases, so if the concentration of hydrogen in Zr was initially near the solubility limit at the lower temperature, the flux due to thermo-transport will raise the concentration at the cold end above the solubility limit and precipitation at the cold end will result. Fig. 7-7 shows initial and final distribution of hydrogen in a dilute zirconium alloy after annealing in a

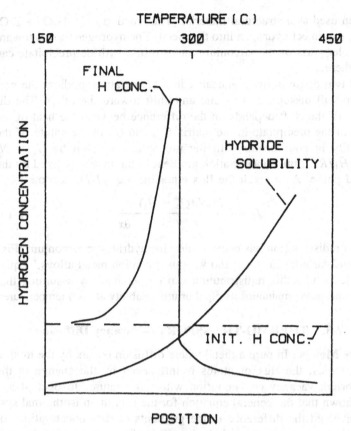

Fig. 7-7—Hydrogen distribution in Zr after annealing an initially single phase alloy in a temperature gradient. [After A. Sawatzky, *Jnl. Nucl. Matl.*, 2 (1960) 450.]

temperature gradient, as well as the solubility of the hydride as a function of temperature. The initial H concentration was above the solubility of H at the cold end of the sample, so a precipitate forms at the cold end, as one would expect. However, the amount of hydrogen found at the cold end after the anneal far exceeds the initial content. This is because most of the hydrogen initially in the higher temperature region has diffused down to form a precipitate at the cold end. In the high temperature region where there is no precipitate, the concentration of hydrogen in solution drops steadily from the solubility limit. This is consistent with a positive Q^*. The H concentration increases rapidly where the solubility of the hydride is exceeded, since any hydrogen swept into this region by the temperature gradient forms a precipitate.

In the core of water cooled nuclear power reactors zirconium alloy is often used as a structural material. The reaction $Zr + H_2O \rightarrow ZrO_2$ + 2H, can inject hydrogen into the metal. The hydrogen moves toward the cooler parts of the zirconium and a brittle hydride precipitate can form there.

If a two phase alloy is annealed in a temperature gradient the precipitate will dissolve at one end and shift toward the other. The direction of the shift depends on the difference between the heat of solution of the precipitate in the matrix $\bar{H}_1$, and Q^* of the solute. If the solubility in equilibrium with the precipitate is given by $N_1 = N_o \exp(-\bar{H}_1/RT)$, the concentration gradient in the matrix is fixed by the second phase. As a result the flux equation, Eq. (7-7), becomes[8]

$$ J_1 = -\frac{D_1 N_1 (Q_1^* - \bar{H}_1)}{RT^2} \frac{\partial T}{\partial x} \tag{7-15} $$

Such redistribution has been studied for hydrides in zirconium (Fig. 7-7), and carbides in iron,[8] and various transition metal alloys.[9] Equations describing this redistribution can be obtained by assuming that the second phase maintains its equilibrium solubility at each temperature.

7-3.   THERMO-TRANSPORT—Vacancy Diffusion

Pure Metals. In pure a metal where diffusion occurs by the motion of vacancies, the flux of atoms is influenced by the change in the equilibrium vacancy concentration with temperature. In Sect. 4-3 it was shown that the general equation for the flux in an isothermal system involved the difference in the gradients of the concentrations of vacancies and atoms. To include the effect of the variation of the concentration of vacancies we start with the equation

$$ J_A = \frac{-D_A}{RT\Omega} \left[-\frac{\partial(\mu_A - \mu_v)}{\partial x} + \frac{Q_A^*}{T} \frac{\partial T}{\partial x} \right] $$

In a dilute solution this can be rewritten

$$ J_A = \frac{-D_A}{\Omega} \left[\frac{\partial ln N_A}{\partial x} - \frac{\partial ln N_v}{\partial x} + \frac{Q_A^*}{RT^2} \frac{\partial T}{\partial x} \right] $$

but if vacancy equilibrium is maintained at each temperature, $(dln N_v/dx)(dx/dT) = -H_v/RT^2$ so

$$ J_A = \frac{-D_A}{\Omega} \left[\frac{\partial ln N_A}{\partial x} + \frac{(Q_A^* - H_v)}{RT^2} \frac{\partial T}{\partial x} \right] = \frac{-D_A}{\Omega} \frac{(Q_A^* - H_v)}{RT^2} \frac{\partial T}{\partial x} \tag{7-16} $$

[8]P. G. Shewmon, *Trans. AIME*, *212* (1958) 642.
[9]Mehmet Uz, D. K. Rehbein, O. N. Carlson, *Met. Trans.A 17A* (1986) 1955–66.

Where the last equality is obtained since in a pure metal the concentration gradient of the matrix atoms is essentially zero. (Here $D_A = D_A^*/f$ as in Eq. (7-6), but the experimental measurement of Q^* is of such low accuracy that we will not carry along the distinction between Q^* and Q^*/f.) This equation could also have been obtained directly from Eq. (7-15) by noting that the local equilibrium concentration of vacancies corresponds to the assumption there of local equilibrium between precipitates and solute in solution.

The net flux of atoms in a pure metal is measured by the motion of inert marker relative to the end of the sample. The net accumulation or removal of atoms on any given plane is obtained by taking the divergence of the flux, Eq. (7-16). This divergence arises from changes in the temperature gradient, and changes in the diffusion coefficient with temperature. Experiments have been done by heating one end a pure metal and cooling the other, or by heating the center of a sample by passing an electric current through it and keeping the ends cold.

Table 7-3 gives experimental values of Q_A^* for several pure metals. Note that Q_A^* is invariably much less than H_v. Thus the vacancy concentration gradient dominates and the net flux of vacancies in always toward the cold end. There are always more vacant sites for an atom to jump into on its higher temperature side than on its lower temperature side. This results in a net flux of atoms up the temperature gradient and an increase in the distance between markers in the hottest parts of the sample.

Since the effect of Q_A^* is small relative to the vacancy gradient effect, Q_A^* cannot be measure with much accuracy. It is not uncommon to have different authors report different signs for Q_A for the same metal.

If one now adds a solute to form an alloy, the process becomes more complex. In this case whether the solute becomes enriched at the hot or cold end of the sample is determined by the competition between the flow of the solvent and solute atoms relative to the lattice.

Table 7-3. Q^* in Pure Metals

Metal	H_m (kJ)	H_v (kJ)	Q^* (kJ)
Ag	81	97	0
Au	86	84	−25
Al	60	63	7
Pb	60	49	9
Pt	132	153	65

Hans Wever, *Elektro- u. Thermo-transport in Metallen*, J. A. Barth, Leipzig (1973), p. 216.

PROBLEMS

7-1. (a) The effective charge for carbon in α-Fe is -12. An electric field of 0.01 V/mm is maintained in a 1 mm long sample of an Fe-C alloy at a temperature of 700° C. What will be the concentration ratio between the two ends at steady-state? (F is 96,500 Coulomb/mole and a Volt-Coulomb = 1 Joule)

   (b) If the initial atom fraction of carbon in the alloy was 0.0005, what fraction of the length of the sample is at a concentration lower than 0.0005 at steady-state, if no carbide forms?

   (c) A carbide forms on the end of the sample where the concentration of carbon is highest whose solubility in the metal is 0.001. For this case what fraction of the length of the sample will be at a concentration lower than 0.0005 at steady-state?

7-2. Referring to Fig. 7-3, if a voltage drop of 0.1 V/cm is maintained in a piece of gold, will the distance between markers at the center (1000° C) and cold end of a sample Fig. 7-4 increase or decrease, and by how much, in 10,000 s? [Take $z^* = -9$ and $D_{Au}^* = 0.04$ cm^2/s exp($-170,000/RT$)]

7-3. Thermo-diffusion arises from the fact that in a temperature gradient atoms jumping from one plane to an adjacent hotter plane do so with a slightly different frequency that that of atoms making the reverse jump. Estimate the ratio of these two jump frequencies using Eq. (7-12) and (7-14), and taking $dT/dx = 500°$ C/cm, $Q^* = 80$ kJ/mol and $T = 1000°$ K.

7-4. If a temperature gradient of 100° C/cm is maintained in a piece of pure gold, will the distance between markers at 900 and 1000° C increase or decrease, and by how much, in 10,000 s? [Take Q_A^* $= -25$ kJ/mol, $H_v = 84$ kJ/mol, and $D_{Au} = 0.04$ cm^2/s exp($-170,000/RT$)]

7-5. Consider the 'thermocouple' shown in Fig. 7-8 with one end at temperature T_1 and the open end at T_2. An interstitial solute is introduced for which $Q^* = -84$ J/mol in one leg and $Q^* = 0$ in the other leg.

   (a) Assuming that the solute concentration is the same in both phases at the interface at temperature T_1, give the equation for the steady-state concentration difference Δc between the two phases at T_2 as a function of $T_1 - T_2$.

   (b) Using this equation calculate Δc if $T_2 = 700°$ C,

$$T_1 = 800° \text{ C} \quad \text{and} \quad c = 1 \quad \text{at} \quad T_1.$$

(The analogy between this system and a normal thermocouple is complete. For example, if solute flows through the junction at T_1, the heat adsorbed or evolved per mole of solute would be just

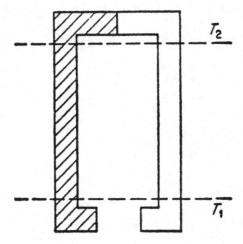

T_2

T_1

Fig. 7-8 —

the difference in the molar heats of solution for the solute in the two phases. The analogous heat in a thermocouple is called the Peltier heat. It is not common to talk about Q^* for the electrons in a metal, but an equation for the voltage $\Delta\phi$ produced at the open end of a thermocouple can be derived which corresponds to the equation for Δc derived above. The equation for $\Delta\phi$ can be expressed in terms of Q^* for the electrons but is more commonly expressed in terms of S^*, the entropy of transport for the electrons where $Q^* = TS^*$. See for example, R. Heikes, R. Ure (eds.), "Thermoelectricity: Science and Engineering," Chap. 1, Interscience Publ. Inc., New York, 1961.)

Answers to Selected Problems

7-1. (a) $J = 0$ so from Eq. (7-2), $ln(N_h/N_l) = -Fz_i^*E/RT$. Thus $N_h/N_l = 4.2$.   (b) 0.77.   (c) 1.0.

7-4. The distance decreases since $(Q_A^* - H_v) < 0$. The difference in $J_A\Omega$ at the two markers is the velocity. The shift is 8.4 μm

INDEX

---

[1] (T) indicates a table of data